Interactivity 2
New media, politics and society

Alec Charles

Peter Lang Oxford

Peter Lang Ltd
International Academic Publishers
52 St Giles, Oxford, OX1 3LU
United Kingdom

www.peterlang.com

This is a fully updated and revised edition of *Interactivity*, 2012
ISBN 978-1-906165-44-4 by the same author.

A catalogue record for this book is available from the British Library.

ISBN 978-1-906165-49-9 (print)
ISBN 978-3-0353-0623-1 (eBook)

Printed in the United Kingdom

Contents

Acknowledgements

Grateful thanks for their support are due to my friends and colleagues Kelly Hallam, Emily Harmer, Peter Harrop, Luke Hockley, Michael Higgins, Dan Jackson, Brendan O'Sullivan, Phil Potter, Bill Rammell, Heather Savigny, Mick Temple, Liesbet van Zoonen, Garry Whannel, Tim Wheeler and Dominic Wring. Thanks are also due to my students, whose feedback and suggestions have been invaluable, and to all those who kindly gave their time to contribute their comments to this study: Jaak Aab, Danah Boyd, Moira Burke, Iain Dale, Gonzalo Frasca, Jane Griffiths, Aleksei Gunter, Joe Hewitt, Andrew Keen, Adam Kramer, Davin Lengyel, Tim Loughton, Austin Mitchell, Mark Oaten, Mart Parve, Larry Sanger and Linnar Viik. Thanks also, of course, to all of my friends and colleagues at the University of Chester. Particular thanks are of course due to Lucy Melville at Peter Lang for her continuing support and for her suggestion that it might be time for a second edition – and also to Alessandra Anzani at Peter Lang for her assistance in the editorial and production processes. And thanks of course to all on Facebook and Twitter.

CHAPTER I

Engines of Change

We should not call the Internet the Internet. We should instead, says James Curran, call it the internet. Curran's point is rather less flippant than it sounds. He observes that nineteenth century liberals had once believed that popular journalism would become an 'autonomous agency of rational and moral instruction' and had therefore capitalized the 'Newspaper Press' and suggests that we have applied the same idealizing or fetishizing attitude to the internet, arguing that it is now time to drop the awestruck capitals and see what this medium is really all about (Curran, Fenton and Freedman 2012: 60). That was the first change upon which I decided when coming to produce a second edition of this book. The second was to do the same thing with *World Wide Web*.

We might also come to interrogate the term 'new media' – which, like Oxford's New College (founded in 1379, but – by just 55 years – the newer of the University's two colleges dedicated to the Virgin Mary), is starting to sound a little old. We will, however, let it stand for the moment, for the purposes of this book, though we might usefully think of the term as being under threat of erasure, or indeed simply in ironic air quotes. The survival of this term is not merely because there is not as yet another phrase in common usage which better serves the intended meaning; it is precisely because the innate inadequacy of the designation 'new media' seems quite appropriate in itself. It conjures the instability and uncertainty of these forms, and also their inherently and interminably emergent, provisional and aspirational qualities. The newness of new media signals a direction of travel. That is both what defines them and what prevents their definition: their refusal to stand still. Our dissatisfaction with the term 'new media' aptly then mirrors their own dissatisfaction with themselves, their constant dynamic imperative to upgrade.

Strange new world

New media technologies have not changed the world. The sun has not fallen from the sky. The rivers still run to the sea. The poor remain with us, and many, many millions continue to toil on the land and to hunger in their beds. We still, for the while, inhabit these vile, frail, mortal shells. While such technologies have clearly made a significant impact upon the ways in which we communicate with each other, we remain corporeal creatures who still eat, sleep, bleed, defecate, procreate and die.

Yet it seems clear that these technologies are radically affecting the ways in which modern societies function. High street stores are closing down, our smartphones oblige us to take our work wherever we go, and nobody writes letters anymore. Bill Gates has killed off the typing pool. As the likes of Keen (2008), Papacharissi (2010) and Gregg (2011) have observed, our daily lives and labours have altered in ways unimagined a decade or so ago. In April 2014 French trade unions reached a legally-binding agreement with employers to prohibit the sending of work-related emails after six o'clock in the evening. It is not so much this agreement in itself, but the need for such an agreement, which may been seen in its own small way to signal a radical change in how mobile information and communication technologies have affected the balance of our daily lives.

When, for instance, we source our music or news or friends online, then the smart technologies which shepherd us through the web anticipate our interests based upon our browsing, networking and purchasing histories and offer us suggestions based on what they predict we might want. This leads us into ever decreasing circles of social, political and cultural exploration: we never grow beyond our original spheres of interest.

The old-fashioned record store, now dying out, would once expose consumers to cultural and social influences beyond their own individual tastes and experiences. It would contribute to the culture and economy of the high street; it would provide retail access to those without the use of online technologies; it would sustain employment in (amongst other areas) retail, manufacturing, construction and design industries.

The speed, ease and convenience of contemporary digital technologies often mean that we fail to become alert to their consequences and costs. These can sometimes prove extremely unexpected, in ways both traumatic and absurd. In April 2014 it was, for example, reported that a British tourist holidaying in South Africa was billed £2609.31 in mobile online roaming fees after downloading a Neil Diamond album onto her iPhone for £8.99. 'It wasn't a particular song that I wanted to hear,' she told the *Daily Mail*: 'I'm really not that big a Neil Diamond fan.'

But have contemporary media forms really made such a radical and permanent impact upon the overall course of human history as some (usually online) might claim? Perhaps not – perhaps not at least in an advertent, directed and constructive way, or in the way we expected and have been told they would. In fact – beyond the hype of the cyber-gurus – the emergent technologies currently most seriously touted as the potential historic game-changers – from genetic modification, gene therapies and nanotechnology to civil nuclear fission, geothermal energy and carbon dioxide removal techniques – all seem resolutely material. These are developments that might not only change our lives but might indeed save our lives and our civilizations.

New media technologies, by contrast, have given teenagers yet another reason not to talk to their parents, twenty-somethings yet another place to meet unsuitable romantic partners, thirty-somethings yet further reasons to regret their twenty-something choices of romantic partners, and the middle-aged still more ways to buy car insurance, world music and garden furniture. It is unclear to what extent late postmodern societies really needed many more of such opportunities. These technologies have given the world an almost inexhaustible supply of ill-informed opinion, hardcore pornography, pictures of kittens, adverts for products and services nobody needs or wants, and video recordings of people doing things at which they are a lot less talented than they seem to think they are – although, again, some might say that even before the irresistible rise of the internet we already had more than enough of these phenomena.

If new media technologies are game-changers, then they are so primarily in the most literal sense of the term: they have changed the games we play, or the ways we play those games, but have not contributed so

much of actually useful substance to the fabric of our material lives. They
have not thus far marked a step change in the evolution of our societies by
revitalizing or inspiring democracies or by promoting political consensus,
civic participation, social inclusion and integration, or the celebration of
cultural diversity. But they have changed the ways in which some people
in some parts of the world buy greetings cards. Even though they still then
tend to send those cards by post.

While it is nevertheless true that these technologies have significantly
enhanced the lives of many physically disabled people in post-industrial
nations in relation to their opportunities for social and cultural access, their
potential even in this area has been undermined by the fact that those able
to access such technologies (especially amongst the elderly and outside the
more affluent regions of developed nations) remain in a minority. This may
of course change over time.

It is obvious enough that we cannot expect all new technologies to
have radical impacts upon the broad sweep of the history of the world. It
is just that in this case many of us thought that these technologies were
going to do so – and that we continue to act as if they have done so or are
still doing so. If the net really did what it said on the tin, then it might be
acceptable for our real-world levels of social and political participation to
diminish – because our online activities would have more than compensated for this. But if new media fail to deliver on their promises, then we
might be losing out precisely because we are unaware we are. We may have
experienced a paradigm shift – but we should bear in mind that a change
in the way we see things is not necessarily the same as a change in the way
things actually are. The map may (for some) be changing, but the world
has not as yet entirely followed suit. That disconnect between perception
and actuality is not unproblematic.

Palfrey and Gasser (2008: 2) have written of a new generation of 'digital natives' – a generation which interacts in ways strange to those outside
this culture – a generation which, they say, will eventually transform global
politics (Palfrey and Gasser 2008: 7). The impact of these technologies
upon society at large may seem clear to such commentators; and yet we
as individuals remain resolutely, physically human. In the first half of the
eighteenth century Jonathan Swift wrote a poem about a besotted lover

stealing into his 'goddess' beloved's dressing room, a lover who, in doing so, comes to recognize in the traces he discovers of her 'dirt, and sweat, and earwax' and of her 'excremental smell' that she is nowhere near as divine as he had imagined. Three centuries on, we are perhaps experiencing something similar: a sense of the absolute absurdity of an existence at once transcendentally virtual and inescapably material – our apotheoses relentlessly deferred by the fact that our technological panaceas cannot ever quite manage to deliver all they promised – caught between two worlds, dynamically suspended in a state of radical transition and of fundamental uncertainty.

Jonathan Swift's poem came to the conclusion that the best solution to this paradox would be to acknowledge and to celebrate its absurdity: that we might, in the end, come to bless our 'ravished sight' for our ability to see 'such gaudy tulips raised from dung.' This, in its own small way, is what this book attempts to do: to expose the paradoxes of our relationships with new media technologies and in doing so to support the possibility of some kind of pragmatic reconciliation.

Reductio ad absurdum

This book is, as the title suggests, about interactivity – or rather it is about the ways in which the declarations or illusions of heightened interactivity advanced by new media technologies and applications offer users a deceptive sense of their own active participation in political and social processes, and thereby undermine those users' desires and potentials for actual, meaningful and impactful engagements in such processes. This book is about that paradox – the way in which a self-avowedly democratizing agglomeration of media forms might effectively counteract and reverse processes of democratization – and therein about the absurdity of that paradox. We may love these new technologies and their uses – *like we don't all use them – like we don't all depend on them* – without that love being irretrievably unconditional. We may be enthusiastic without being unquestioningly fanatical; we may rely upon them without putting our absolute trust and faith in them, in what they are in themselves and excluding all else. There is clearly nothing fundamentally wrong with emergent communication and information technologies: technology in itself is neither nemesis nor panacea; it is at best merely a facilitator. The problems arise when we come to see it as a solution in itself. That is what this book is about.

This book is also therefore about the absurdity – the definitively paradoxical nature – of contemporary existence, existence, that is, in the information society of the post-industrial world, a domain of irreconcilably clashing paradigms: it is about what happens when, for example, as Jesse Rice (2009: 21) puts it, 'our brilliant but imperfect humanity collides headlong with Facebook's brilliant but imperfect technology.' The twentieth century French philosopher Albert Camus (1975: 13) once similarly identified the feeling of absurdity as resulting from a fundamental existential disconnection, suggesting that a 'divorce between man and his life [...] is properly the feeling of absurdity.'

This then is the extent of the absurdity we face, an absurdity modelled and thereby aggravated by the forms and genres of contemporary mass media culture. The virtual society – this civilization of electronic government, digital play, reality television and online networking – generates a world in which the empty fantasies of *World of Warcraft* parallel the prevailing condition of absurdity experienced by a civilization which has sought to wage a War on Terror, an absurdly abstract and devastatingly immediate phenomenon which keeps refusing to be consigned to the pages of history. This book explores how these media forms perpetuate and mirror this absurdity, and suggests that, although we may claim not to have started the fire, it appears we have been attempting to put it out with gasoline.

The absurdist short-story writer Jorge Luis Borges (1970: 42) famously imagined the fantasy world of Tlön, a world conceived as a theoretical model by intellectuals, a virtual world whose structures come to overwhelm the real world: 'How could one do other than to submit to Tlön, to the minute and vast evidence of an orderly planet? It is useless to answer that reality is also orderly. Perhaps it is, but in accordance with divine laws – I translate: inhuman laws – which we never quite grasp. Tlön is surely a labyrinth, but it is a labyrinth devised by men, a labyrinth destined to be deciphered by men.' We have invented a new world to replace the incomprehensible cosmos created by a divine being, and that world has become the reality which we now inhabit. Yet this artificial universe, though absolute and incontrovertible, remains unintelligible to us. This is perhaps the continuing absurdity of the human condition – a double absurdity in that the paradigm developed to overcome our existential absurdity has proven even more paradoxical, even more absurd, than the obsolescent theocentric perspective it was designed to supersede. In order to bridge Camus's divorce between subjectivity and existence, we have constructed an anthropocentrically bespoke and irresistible mode of being – one from which we, nevertheless, feel even more alienated than ever before. There yet remain defining and perhaps unerasable traces of our abandoned paradigms, and the inconsistencies and contradictions between these and the perspectives which we are now adopting propagate a sense of inevitable and almost inconsolable absurdity.

Almost inconsolable – this absurdity is only *almost* inconsolable insofar as, as Camus (1975: 53) proposed, the solution to this dilemma is to recognize the inevitability of that absurdity and to accept the impossibility of transcendental meaning: 'life [...] will be lived all the better if it has no meaning.' To become reconciled to the irreconcilability of existence is, for Camus (1975: 63), to embrace 'the absurd world [...] in all its splendour and diversity.' And yet, because this is clearly nearly impossible, we appear to return time and again to all of those artifices (e-democracy, video games, social media, reality TV) which pledge to banish the absurdity and to instal each one of us in our rightful place as the controlling and beloved agent, literally the *avatar*, the divinity incarnate, of our existence, and which, in doing so, only serve to exacerbate the extent of our existential alienation.

Youtopia

Zizi Papacharissi (2010: 3) has suggested that new communicational technologies tend to provoke 'narratives of emancipation' which are framed between 'utopian and dystopian polarities.' The cyberutopianists, cyberenthusiasts or cyberoptimists, as they are sometimes called, tend to suppose that emergent information and communications technologies offer a panacea for all of the ills of the world and thus herald an electronic utopia of democratic, economic and educational freedoms and rights on a global scale – advancing a 'technotopia' (Storsul and Stuedahl 2007: 10) or 'a New Alexandria' (Koskinen 2007: 117). As Tiffin and Rajasingham (2003: 26) have suggested, the internet has been seen by some as 'a democratic virtual meeting place [...] a new Museion of Alexandria.' Indeed in the earlier days of mass internet use Tiffin and Rajasingham (1995: 7) had gone rather further in their enthusiasm for the potential of this technology: 'VR offers us the possibility of a class meeting in the Amazon Forest or on top of Mount Everest; it could allow us to expand our viewpoint to see the solar system operating like a game of marbles in front of us, or to shrink it so that we can walk through an atomic structure as though it was a sculpture in a park; we could enter a fictional virtual reality in the persona of a character in a play, or a non-fictional virtual reality to accompany a surgeon in an exploration at the micro-level of the human body.' One is tempted to recall in this context the avowed cyber-realist Andrew Keen's scepticism as to a 'mad utopian faith in our ability to conquer the physical world through virtual reality' (Keen 2008: xviii).

Some have lauded new media technologies as the key to the global establishment of liberal democracy, often without recognizing the limitations of these tools, their potential for appropriation by reactionary and totalitarian institutions of power, or for that matter the questionable universal appropriateness and value of such a political system. J. Hunter Price (2010: 1) has, for example, pointed out that, as these technologies have been increasingly deployed by opposition campaigners in dictatorial states, western commentators have grown 'even more confident in their conviction that new media will lead to a wave of democratization.' This is the dream of the cyberoptimists. As Joss Hands (2011) has declared, @ is for Activism.

The cyberpessimists, by contrast, tend to suppose that George Orwell was right: that we are increasingly living in a surveillance society, a hegemonized and homogenized realm, a media dictatorship of totalizing ideologies and corporate hyperpower – the world which Andrew Keen (2008: 166) has dubbed '1984, version 2.0.' This is a mode of Orwellian dystopia which is as consensual as it is totalitarian: as Keen (2008: 175) reminds us, 'the age of surveillance is not just being imposed from above by aggregators of data. It's also being driven from below by our own self-broadcasting obsession.' This is the YouTube youtopia.

This is a world in which new media technologies sponsor fanaticism and terrorism as much as popular liberation. This is a world in which the notion that the internet empowers the powerless is, as Morozov (2011a: xii) suggests, no more than so much 'cyber-utopianism'. The cyberpessimistic position tends to be epitomized by anxieties as to what Talbot (2007: 173) has identified as the 'increasing corporate control of the internet.' Such reports as that of the 2010 agreement between search engine Google and Verizon (America's largest telecommunications company), described by *The Guardian* newspaper (5 August 2010) as 'a deal that could bring an end to net neutrality', have led to fears of a double speed internet that would reinforce digital divide (Frau-Meigs 2007: 48) and undermine the democratic potential for which (in everything from electronic government to the social networking site) the internet has been celebrated by the cyberoptimists. It was reported that Google and Verizon spent a combined total of $63 million on political lobbying between 2012 and 2013. In early 2014 various online commentators – including FireCable's Steve Donohue – suggested that the two companies might even be considering a merger.

In April 2014 Google – along with Apple, Adobe and Intel – agreed to settle an antitrust lawsuit out of court – a class action which had claimed that the companies had colluded in their hiring practices to avoid poaching each other's staff. Google had also come under fire in March 2012 when the European Union's Justice Commissioner Viviane Reding declared that changes to Google's privacy policy – changes which allowed the search engine to share user data with its other platforms (including Gmail and YouTube) – breached EU law. Her announcement not only prompted the *BBC News* website to offer its readers instructions as to how to delete their Google histories, but also echoed and amplified concerns that have increasingly dogged the corporate giants of the internet industry. (Cf. Bell 2014.)

In August 2012 it was reported that Google had agreed to pay the largest fine ever imposed on a company by America's Federal Trade Commission.

The $22.5 million fine was imposed after it had been discovered that Google had been monitoring users of Apple's Safari browser who had enabled the 'do not track' privacy setting. In February 2013 it was reported that the European Union was planning to take further action against Google's strategy of combining data from across its various platforms in order to enhance its targeting of advertising, a strategy which was considered to put users' rights to privacy at risk.

The following month Google agreed to pay a further $7 million fine for harvesting personal data from home wireless networks without permission during the operation of its Street View service. As part of the settlement, the company also agreed to delete the harvested emails, passwords and web histories. One might hesitate to speculate as to whether the decision-makers at Google consider that the company is convincingly continuing to live up to its almost utopian corporate statement of core values: 'don't be evil' – 'we make money by doing good things.'

In April 2014, amidst further privacy concerns, Google reversed a policy to scan accounts linked to educational institutions for the purposes of advert targeting. Glancy (2014) has again raised privacy concerns in relation to the extensive but usually unread contracts of usage required by online corporations and in particular 'the ever-fluid, ever-mystifying terms of Google and Facebook.' The problem, stated simply, is this: in the world of the web there is no such thing as a free search (or a free social network or host for your emails, videos or blogs, for that matter). Organisations like Google and Facebook rely on advertising in order to generate revenues essential not only to sustain their burgeoning corporate and technological infrastructures but also to reward their shareholders. In May 2012 Facebook launched its first stock market flotation, offering $18 billion dollars' worth of shares – the result of the highest valuation of internet stock in an initial public offering since Google had raised $1.67 billion dollars in 2004, but still only a fraction of the estimated $104 billion valuation of Mark Zuckerberg's company. As the users of these services remain unwilling to pay for them themselves, these organisations have therefore looked to advertisers to meet the costs not only of development and delivery but also of dividends. What advertisers want from these companies is targeted access to their hundreds of millions of customers, based upon data harvested as to these customers' online preferences and personal information. Our value to Google and Facebook is our private data. (Cf. Waters 2014.)

This is why one's Facebook page features adverts for products for which one has indicated a fondness in one's lists of likes. This is why some sites deny access to computers set to refuse the temptations of those innocently named 'cookies' – packets of data which track and hold information on user preferences. Research published by the online privacy solutions organization Truste in April 2012 suggested that users encounter an average of 14 such trackers per page on Britain's most popular websites and that 68 per cent of these trackers belong to third-party organizations. In Europe such has been the concern as to the intrusiveness of cookies upon users' privacy that new laws came into force in May 2012 obliging websites to seek users' consent before harvesting data via cookies – although in the UK the Information Commissioner's Office warned of widespread non-compliance with the terms of this legislation. Two years on, it is still common enough to find websites which neglect to request such permissions.

In March 2012 – in response to the Google privacy controversy – the director of the civil liberties group Big Brother Watch told the BBC that he feared that Google was now 'putting advertisers' interests before user privacy.' In August 2010 Google boss Eric Schmidt had informed *The Wall Street Journal* that he did not believe that privacy regulation was necessary – because his users could abandon the site if Google did anything with their personal information that they found 'creepy'. But do internet users really have much of an alternative? Indeed one might ask whether they might have the right to demand higher ethical standards from these self-proclaimingly idealistic companies – when, as Fuchs (2011) has suggested, the net's most popular platforms allow their users scant opportunities for participation in the development of their own terms of use policies.

Around the turn of the millennium the architects of Web 2.0 had heralded the online age in democratic, communitarian and egalitarian terms which promised the technotopia of a global village. So where did it all go wrong? Today's moguls of the net have this much in common with the leaders of any revolutionary movement: they have discovered that the maintenance of power, once achieved, all too easily seeks recourse to the established hierarchies, structures and interests which their idealism had once sought to overthrow.

Yet the optimism of many cyberenthusiasts continues apparently unabated. The optimists or cyberenthusiasts represent a position once summed up by Bolter and Grusin (2000: 59–60) in the following terms:

> They tell us, for example, that when broadcast television becomes interactive digital television, it will motivate and liberate viewers as never before [...] that hypertext brings interactivity to the novel [...] that the World Wide Web [...] can reform democracy by lending immediacy to the process of making decisions.

This enduring position has been neatly challenged by Koskinen (2007: 125):

> What is the point in selling the idea that digital TV makes it easier for us to order pizza when any modern city already provides plenty of opportunities for ordering pizza? [...] Take the notion of interactive narratives [...] No one in his right mind can write an alternative ending to the story of Jesus Christ. Or what is the point in taking *Romeo and Juliet* and attempting to improve its dialogue by making it interactive?

James Curran (2012: 34) has observed that early histories of the internet 'were conditioned by the awestruck period in which they were written. Their central theme is that utopian dreams [...] led to the building of a transformative technology that built a better world.' But, as Liesbet van Zoonen (2010) has suggested, although there has been a lot of utopian talk about what Web 2.0 can do, its failure to achieve those socio-political ambitions has been quite embarrassing. Papacharissi (2010: 20) points out that processes of commercialization have simply turned new media into versions of older media – 'expanding shopping catalogs for the consumer, but not affording democratic options for the citizen.' The internet might then from this perspective come to seem no more than a domain of cynical commercialism, of user narcissism and paranoia, of pornography and price comparison websites, of cameraphone selfies and videos of adolescent drinking stunts.

Martin Hand (2008: 15) has suggested that the digital age has been greeted with great optimism by its enthusiasts and denounced with equal passion by its detractors:

Most commonly, narratives of digital culture imbricate western models of democratization with enthusiastic accounts of information technologies. For some, such technology is instrumental in broader restructurings of modern society, replacing structure with flow, state with network, hierarchical knowledge with horizontal information [...] For others, the use of the term [...] 'digital culture' is hasty or simple determinism, reifying either information or technology as great levellers, an ideological rhetoric which has the effect of glossing an increased penetration and 'hardening' of global capitalism.

However, despite the apparently irreconcilable nature of the cyberoptimistic and cyberpessimistic stances, we might concur with both the cyberoptimists and the cyberpessimists on one central theme: that the evolution and adoption of new media technologies heralds a paradigm shift in our notions of politics, society and subjectivity. This is not, of course, the end of civilization as we know it, but a transformation of our model of civilization. This transitional period defies rationalization in traditional terms, generating instead paradoxes whose absurdity reveals an impasse within received notions of identity, meaning and societal progression.

This book does not suggest that this is a bad thing *per se*, but proposes that our time-honoured theories and ideals of ideological self-determination, social interaction and representative democracy are becoming increasingly irrelevant to the conditions we are heading towards: that these perspectives are coming to represent what Papacharissi (2010: 8) has called 'metaphors that no longer work.' This is not the critical apocalypse of nuclear armageddon, of fundamentalist terrorism or of anarchic revolution; this seems an ongoing repositioning of history itself, the propagation of a view of a world seen as being without material history, a virtual gameworld without an immediate awareness of material commitment. Like global warming, the future extent of this process is (thus far) almost undetectable to the naked eye. The transfiguration of the western cultural paradigm takes place virtually unnoticed, save for that mild feeling of discomfort – of metaphysical alienation or absurdity – of which some complain. And so the frog boils. This is how T. S. Eliot (1925) famously said the world would end – *not with a bang but a whimper.*

Phubbed Off

A graphically convenient way to demonstrate the ongoing paradigm shift from the traditional material-historical perspective towards a homogeneous, mass-mediated, globalized world view is to conduct a Google image search on the word *Homer*. The overwhelming majority of the images generated by the search will show Bart Simpson's father, rather than the legendary Greek poet and originator of western literary civilization. This prioritization of immediately contemporary, homogeneously popular culture is evidenced by a variety of more formal studies. In August 2006 the BBC reported that a survey commissioned by an online game show devoted to modern popular culture had shown that more Americans knew who Harry Potter was than Tony Blair. While 57 per cent of the 1,213 participants surveyed recognized Harry Potter, less than half recognized the UK's then Prime Minister. The survey also showed that, while six out of ten people knew Homer Simpson's son was called Bart, only a fifth could name one of Homer's ancient Greek epics.

A sense of history (in its traditional sense) is diminishing. In December 2008 it was reported that, in the year of his election to the U. S. Presidency, Barack Obama trailed in third place in the league of Yahoo's greatest number of searches behind pop singer Britney Spears and World Wrestling Entertainment. In June 2011, it was reported that Obama had gained third place in Twitter's most followed tweeters: this time beaten by teen pop sensation Justin Bieber and surreal pop sensation Lady Gaga. Political history comes in as runner-up to populist mass media culture. In November 2009 *The Daily Telegraph* informed its readers that a survey had discovered that 5 per cent of British schoolchildren believed that Adolf Hitler had been the coach of the German football team – while one in six youngsters believed that Auschwitz was a Second World War theme park and one in 20 thought the Holocaust had been a party to celebrate the end of the war.

A report published in August 2010 by Britain's Office of Communications announced that the British were spending almost half of their waking hours watching television or using mobile phones and other communication devices. The report added that the British were also immersed in several types of media at the same time – on average fitting nearly nine hours of daily media engagement into just over seven hours. Research into habitual internet use has demonstrated that online activities are altering the very structures of our brains: Kanai et al. (2011) have, for example, suggested a direct correlation between the number of friends held on social networking sites and the density of brain matter in the right superior temporal sulcus, left middle temporal gyrus and entorhinal cortex.

Andrew Keen (2008: 160) has noted a study by Stanford University in which one in eight adults were shown to manifest symptoms of internet addiction. In November 2009, the BBC reported that a British schools organization had listed SNS addiction as the primary cause for parents' concern about their children's welfare, suggesting that some children appeared to have become 'permanently connected' to such sites as Facebook. In May 2010 a survey conducted by UK Online Measurement suggested that British web users were spending 65 per cent more time online than they had just three years earlier. The survey added that the largest single portion of that time (22.7 per cent) was spent on social networking or blogging sites (as opposed, say, to e-mail use which came in second at only 7.2 per cent of users' time).

A Dutch website called Web 2.0 Suicide Machine has been set up to allow people to disconnect from – and delete their presence within – their online social networks. Its founder told the BBC in September 2010 that his clients 'are basically getting their analogue life back.' However, a psychiatrist who (according to the BBC) 'treats patients who use the internet excessively' warned that 'by disconnecting you are losing a significant relationship. Those 30 or 40 hours of time now have to be filled with real life.'

Gary Small, a researcher into internet addiction at UCLA, had told *The Guardian* newspaper the previous month that 'the internet lures us. Our brains become addicted to it. And we have to be aware of that, and not let it control us.' In April 2011 a report in *The Daily Telegraph* cited research conducted by the University of Maryland's International Center for Media and the Public Agenda which found students across the world admitted being so 'addicted' to their mobile phones, laptops and social networking sites that 80 per cent suffered significant physical and emotional distress when stripped of these technologies for just one day.

In May 2012 researchers at Norway's University of Bergen published a study which likened forms of internet dependency to drug and alcohol addiction, echoing the results of a Chinese study which had been published just four months earlier. In January 2014 it was reported that China had established treatment camps to cure afflicted teenagers of internet addiction. The same month *The Daily Telegraph* reported on the findings of British research which showed that 'workaholics are increasingly logging on after work, becoming addicted to the web.' (Cf. Fisk 2014.)

Andrew Keen has argued that access to the internet is 'something that many people would struggle to live without.' He has added that 'mobile technology makes it more addictive' and that 'if you haven't been on it for a few hours, you get withdrawal symptoms.' If this is a form of addiction then it is one which can display particularly unpalatable manifestations. In April 2012, for example, a cross-party report from the UK parliament revealed that 80 per cent of 16-year-olds regularly access pornography online – and that a third of 10-year-olds have also done so; or, as the *Daily Mail* put it, a 'generation of children is growing up addicted to hardcore internet pornography.'

Some much, then, for romance. In April 2014 the British guerrilla artist Banksy launched a work showing a couple engaged in a night-time romantic embrace against a backdrop of stars. Their faces are illumined not by the moonlight, however, but by the glow of their mobile phones which each is checking behind the other's back. That same month, Helen Clements and her two children had been caught in a burning car in the middle of the lion enclosure at a British safari park. After their rescue, she had commented: 'My daughter was mainly upset that she'd left her phone in the car.'

A year earlier, in April 2013, the BBC had reported that research conducted by Nottingham Trent University had shown that using social media to interact with friends and family on social media 'did not improve relationships.' On 4 August 2013 the BBC denounced 'the latest in modern-day bad manners, people texting and checking the internet on their mobile phones while ignoring the people they are actually with [...] a phenomenon called "phubbing" – snubbing someone in a social setting by looking at your phone.' The following month *The Daily Telegraph* reported on the findings of a study which 'found that mobiles, far from connecting people, break up their lives.' The problem, the paper commented, was that the mobile telephone had become, for many, 'a kind of imaginary friend.' The next month a survey published by the Internet Advertising Bureau revealed that the average British person uses some form of internet-connected device 34 times a day. On 17 October Sky News reported that 'those who took part in the study averaged a total of two hours and 12 minutes a day using a connected device, while for 46% of this time they were using at least two devices, and sometimes three, simultaneously. More than half of mobile phone users (52%) said they prefer to check it if they have any "downtime" rather than sit and think. The figure rises to 62% among 18 to 30-year-olds. More than one third (37%) said they check their phone if there is a lull in conversation with friends.'

Exactly one week later the BBC reported the case of those 'Japanese men who prefer virtual girls to sex' – men whose 'girlfriends' are in actuality small portable tablets bearing the Nintendo computer game Love Plus – men who take these imaginary consorts 'on actual dates to the park, and buy them cakes to celebrate their birthdays.' The BBC quoted one user of Love Plus: 'I'll continue the relationship forever,' the devoted 39-year-old said.

This phenomenon is not of course exclusive to Nintendo or to the Japanese. On Valentine's Day 2013 BBC technology reporter Dave Lee told of how he had 'paid for make-believe love on Facebook.' Lee had talked to an online escort who, for a $5 fee, would become her client's online girl-friend for a week, including as part of the service a few appropriate status updates and likes: 'It's not a big deal really,' she told the BBC. 'I just tick *in a relationship.*'

On 7 January that year the BBC had quoted Mark Griffiths, an addic-tions expert at Nottingham Trent University, who had suggested that video gaming habits should not generally be considered an addiction: 'I've come across very excessive players – playing for 10 to 14 hours a day – but for a lot of these people it causes no detrimental problems if they are not employed, aren't in relationships and don't have children.' On the assumption then that work, family and real-world human relationships are no longer con-sidered priority lifestyle choices, it seems that there is very little in these developments for society to be greatly concerned about.

Greenfield (2013) has thus written of the emergence of a new infantilism: 'a generation of 20-somethings still living at home, wearing "onesies" [...] perhaps playing mythical or sci-fi games [...] and/or craving the constant attention of others through social networking sites.' Is this then the shape of things to come, generation upon generation of Twitter-Tweenies till the end of time (or at least until bed-time)?

The immaterial world

As electronic media move our perspectives beyond the traditional constraints of space and time, and restructure the ways we think and perceive, the phenomena of existence appear to be becoming increasingly immaterial. This has been going on for quite some time. It was, in fact, about the time that Madonna announced we were living in a material world that we started to realize we were no longer were. That was in 1984, but even George Orwell could not have seen what was coming. The Internet Activities Board had been established just the previous year. The virtual world had begun.

Welcome, therefore, to *the desert of the real*. This sentiment was advanced by that great prophet of the virtual world, the sociologist Jean Baudrillard, in one of his last books (Baudrillard 2005: 27). In deploying this phrase Baudrillard was alluding at once to Slovenian philosopher Slavoj Žižek's book of 2002, *Welcome to the Desert of the Real*, and to that line of Laurence Fishburne's from the 1999 film *The Matrix* – from which Žižek's book takes its title. The Wachowski brothers' film – which features a copy of Baudrillard's earlier work *Simulacra and Simulation* – itself takes that phrase, *the desert of the real*, from the very first page of that book, Baudrillard's own seminal work (Baudrillard 1994: 1). This most postmodern cycle of referentiality offers a pertinent metaphor for the contemporary condition of historical disorientation. A sense of originality and authenticity has been lost within this post-material reality – as is witnessed in this allusion to an allusion to an allusion to one's own original object (Baudrillard's allusion to Žižek's allusion to the Wachowskis' allusion to

Baudrillard). The allocation of provenance (and therefore the possibility of historical logic) becomes unfeasible in a cultural environment dominated and epitomised by this recurrence of reflection, this palimpsest of simulacra (copies without originals, shadows without substances, reflections without objects). The desert of the real is not, as Laurence Fishburne's Morpheus supposed, the material world: it is the meaningless void of the virtual realm, the world that most of us in the western world spend so much of our time inhabiting. Yet, although there is no authenticity, we are still tormented by a nostalgic yearning for it. Without authenticity as its opposite, there can be no true fakery; yet our desire for authenticity necessitates that fakery (if we cannot have the real thing, we feel the need to fake it). This is the paradox and the tragedy of the simulacrum: the realm of the copy of the copy of the copy, those relentlessly diminishing reflections which have lost any memory of their original object, *mise en abyme*.

In the mid-1990s Baudrillard (1995: 85) had imagined that the new world order would be a hypermediated one. By the time of his death in March 2007 it appeared that his prophecy had been fulfilled. The world, as he came to see it, was transcendentally mediated: real-time media had come to define real time, reality television had determined a televisual reality – a reality which was at once perfect, totalizing, virtual, integral, utopian and hyperreal, a post-material ultra-reality (Baudrillard 2005: 27). Baudrillard (2005: 32) witnessed an eventual paradigm shift beyond the simulacrum – 'that which hides the absence of truth' – into a reality which now acknowledges (and indeed takes for granted) truth's absence.

Baudrillard depicts a world without distance or difference, a world beyond history and memory, in which the media can no longer distinguish between existence and representation, a world which has witnessed a blurring of the distinctions between democracy and television, between war and games, between society and virtuality, between fact and fantasy. We need look no further than the much-reported case of David Pollard – who in 2008 left his wife (whom he had married in Second Life) for a woman he had met in Second Life – and with whom his wife caught him cheating in Second Life – to marry his new partner both in Second Life and in real life – to see that the ontological prioritization of the virtual world is already, for some at least, a *fait accompli*.

The processes of mediation have become at once so hegemonic and so augmented that it is increasingly difficult to tell what is real and what is not – insofar as the real itself is defined by those processes of mediation. Reality television has for instance become so absurd, so beyond any sense of what might once have been perceived as realistic, that one may be forced to conclude that it only maintains its epithet by virtue of the fact that it is now such televisual phenomena which determine how we measure and recognise reality in the external world. Having appropriated and shifted the benchmarks for normative reality, reality television programme formats and practices have, in their attempts to court audiences immune to the appeal of the merely ordinarily extraordinary, grown ever more eccentric. In January 2005, for example, Fox Television had launched a short-lived reality TV show entitled *Who's The Daddy*, a programme in which adopted children would stand to win $100,000 by guessing the identities of their biological fathers from a line-up of eight men. Two months later German TV channel RTL II announced that its new series of *Big Brother* would feature not just a house but an entire village and would have no end date set. (One has flashes of Patrick McGoohan's 1960s fantasy series *The Prisoner*.) The show was cancelled the following February as a result of poor ratings. In both of these cases we may assume that the failures of these programmes suggest that their creators were pitching somewhat ahead of their audience's horizon of expectations. A decade on, those horizons seem somewhat unambitious.

Reality television has continued to push its own boundaries. September 2007 saw CBS's launch of *Kid Nation*, a series in which 40 children (some as young as eight years old) were left to fend for themselves for 40 days in a ghost town in the New Mexico desert. (One has flashes of *Lord of the Flies* – as directed by Sergio Leone.) Others have gone even further: in September 2007 the presenter of a Brazilian true-crime show was accused of ordering contract killings in order to boost his programme's ratings.

Contemporary television is becoming, it appears, almost an inadvertent parody of itself. A prime example of this shift into the realm of the absurd might be *The Naked Office*, launched by the UK's Virgin television in July 2009, a show in which a marketing company in Newcastle attempted to boost morale and productivity by introducing Naked Fridays – or, for that matter, *Autistic Superstars*, screened by BBC Three in May 2010, a well-meaning talent contest for young people with autism – or indeed in December 2011 when Valerio Zeno and Dennis Storm, the presenters of Dutch television's *Guinea Pigs*, ate (surgically removed and sautéed) sections of each other's flesh. Populist television seems to be consuming and regurgitating itself with the frenzy of a cannibalistic bacchanal. In October 2013 Britain's Channel 4 launched *Sex Box*, a series in which couples would have sex inside a large metallic box in the TV studio, only to be quizzed by a panel of judges after they emerged sated therefrom.

When, however, reality television intentionally caricatures itself, the willingness of the public to accept the parody at face value further demonstrates the extremes and absurdities which this mediation of reality has already reached. We have increasingly come to accept the most absurd caricatures of reality television as real, and a number of hoax reality television shows have illustrated this phenomenon. In December 2005 Britain's Channel 4 launched a series called *Space Cadets* in which nine people believed they were in training in Russia to become cosmonauts (rather than, as it turned out, still in the UK on a mock training programme) and that the winner would be launched into space. In May 2007, in order to highlight a shortage of organ donors, Dutch television caused an international controversy when it screened *The Big Donor Show* – in which an apparently dying woman (in reality played by an actress) would choose from among three contestants (real patients in need of transplants) to receive her kidneys. In March 2010 French television faked a reality television show in which 80 people were asked to zap a contestant with increasingly powerful electric shocks each time he failed to answer a question correctly – and in which, urged on by the cheers and cries of 'punishment' from the studio audience, 64 of these 80 members of the general public, despite the contestant's screams and pleas for mercy, pushed the punishment to what they believed to be a potentially lethal voltage.

Within the depthless environment of early twenty-first century mass media culture, these hoaxes (however eccentric, however far they pushed the limits of rationality and taste) were not only believable – they were actually believed – not only by their participants (as in the case of the British and French programmes) but also (in the case of the Dutch programme) by the most seasoned and sceptical media commentators who – prior to its broadcast – lined up to call for the cancellation of the show, their protests framed within a credulous rhetoric which repeatedly invoked the lamentable state of contemporary popular culture. Indeed, this author himself appeared on air that year to denounce *The Big Donor Show*.

Yet the unbelievable absurdity of such programmes as *Space Cadets* has now begun to be realized: not to be recognized, but to be made real, in all its absurdity. History is repeating itself: first as farce, then as fact. In March 2014 *X Factor* front man Dermot O'Leary presented *Live from Space* for the UK's Channel 4, a live broadcast in which viewers could 'interact with the astronauts onboard the International Space Station.' In August 2013 it was announced that one of the creators of *Big Brother* was developing a 2023 television series in which participants would take a one-way trip to establish the first human colony on the planet Mars. The cost of the project had been estimated at £4 billion and 165 thousand people had applied to take part.

Terms of estrangement

It was reported in July 2011 that for the first time the computer technology and digital media corporation Apple had more cash reserves than the United States government. Figures from the U. S. Treasury Department had shown an operating cash balance of $73.7 billion, while Apple's most recent financial results had boasted reserves of $76.4 billion. The following month it was announced that Apple had become America's most valuable company, with its market capitalization for the first time surpassing that of oil giant Exxon. The iPad had become bigger than the oil rig, the iPhone more robust than the United States Treasury. That a media organization could be richer than the world's richest nation, or for that matter than that traditionally oil-reliant nation's biggest oil producer, seems absurd; but it sometimes seems that the twenty-first century's post-industrial civilization can appear so consistently absurd that we barely any longer even notice it. Indeed in the first quarter of 2014 alone Apple reported revenues of $45.6 billion – which, by a simple extrapolation, would put its projected annual revenues somewhere between the GDPs of New Zealand and Kuwait.

New media appear at times to make individuals and organisations lose all sense of themselves, and of proportion. We might select any number of media-related news stories to illustrate this phenomenon ... the accounts, for example, in May 2010 of 'riots' erupting during delicate negotiations between British Airways and the trade union Unite after the union's General Secretary had leaked details of the meeting onto Twitter. Or the tale that (in March 2011) a Blackpool councillor had described his constituents as 'donkey-botherers' on Facebook, or the following month that a Kent Conservative candidate had branded local women 'sluts' on Facebook – or the reports of a Liberal Democrat councillor who, in February 2009, had faced calls for his resignation after posting onto Facebook pictures of himself dressed in Nazi uniform. Or the news of February 2010 that the toy manufacturer Mattel was planning to release 'Puppy Tweets', a device to allow dogs to post on Twitter. Or the epic narrative of a Yorkshireman concerned that heavy snowfall at his local airport would spoil his travel plans who was arrested under the Terrorism Act, suspended from work, banned from the airport and obliged to wait 30 months to be vindicated by a high court ruling, after he had tweeted in January 2010: 'You've got a week and a bit to get your shit together, otherwise I'm blowing the airport sky high!' Or the similar case (of January 2012) of the British man who was refused entry into the United States after having tweeted that he was going to 'destroy America' (the BBC report observed that 'he insisted he was referring to simply having a good time'). Or the instance of an American family whose house was raided by an anti-terrorism taskforce in July 2013 after having made the mistake of Googling for information about pressure cookers and backpacks. Or the revelation in May 2012 of the sorry saga of the British Prime Minister who had signed off text messages to one of the country's leading media executives with the abbreviation 'LOL' believing that it meant 'lots of love' rather than 'laugh out loud' – an innocence reminiscent of those who believe 'WTF' means 'well, that's fantastic.'

In September 2002, when a woman called Rena Salmon murdered her husband's lover and then texted him to tell him she had done so, she added in that message that her statement was 'not a joke.' Williams (2003) has witnessed in this assertion the failure of contemporary society to be alert to authenticity – which is perhaps to suggest that (when, for example, acts of war come to look like Hollywood movies or video games) we have become so far removed in our daily lives from traditional verisimilitude that even in the moments of the most urgently immediate reality we doubt the veracity of material existence: 'that surely *cannot* be a real plane hitting that skyscraper,' news viewers might have mused, 'that *must* be a computer-generated effect.' By translating it into the medium of the text message or the tweet, the event becomes at once real and unreal: it enters into the new paradigm of a hypermediated reality, which is at once detached from material reality (and therefore from traditional moral responsibility) and is also the only reality there now is.

Just as these modes of mediation farcically frame our lives, so they tragically frame our deaths. In October 2007 Anthony Henderson was gaoled for three years after having been filmed on a mobile phone urinating on a woman as she lay dying in the street and shouting 'This is YouTube material!' In March 2010 it was reported that a South Korean couple had become so obsessed with nurturing a virtual daughter in the popular role-playing game Prius Online that they allowed their own three-month-old baby starve to death. The following month Minnesotan William Melchert-Dinkel was accused in court of encouraging the suicides of a British man and a Canadian woman by posing as a female nurse in online suicide chatrooms and advising people on how to take their own lives.

The convergence of new media forms and technologies (reality television, social networking websites, the home video camera) foster unprecedented conditions of cultural and social interaction whose rules of negotiation are as yet poorly understood and which may as a result catalyze unprecedentedly horrific and terrifying possibilities. In October 2009 Petros Williams made a 'farewell video' of his four-year-old daughter and two-year-old son as they watched *The X Factor* on television before strangling them with internet cables. When his case came to court the following March prosecutors described his crimes as representing a 'symbolic act of punishment' against the children's mother for her use of internet dating sites to converse with other men. *The Sun* newspaper reported on 23 March 2010 that the mother had moved out of the family home shortly before the killings as a result of 'rows over her use of the web.'

The fact that we cannot make any sense of this does not make it any the less real. Indeed the reason why we cannot make any sense of this at all is precisely the fact that it is real – but it does not fit within our traditional paradigms, the outdated paradigms of a pre-globalized, pre-virtualized (still heterogeneous, still material) world – nor, conversely, within the expectations of a hyper-controlled utopian virtuality. The absurdity is not our current situation *per se* – the absurdity is that (because we still recognize traditional notions of society, politics and material history) we continue to try to apply our old logic to our newly imagined environment.

Burgeoning absurdities

Through the exponential developments in their spread and speed of operation, new media at once authenticate and exacerbate a contemporary state of absurdity, transforming implausible fantasy into accepted reality. Stories of entirely imaginary provenance have thus repeatedly appeared in the output of established media institutions. For example, in September 2009 two Bangladeshi newspapers published an article taken from the satirical American website *The Onion* which claimed that Neil Armstrong had told a news conference that the moon landings had been faked. In August 2011 it was revealed that reports from the BBC, CNN, *The Daily Telegraph* and the *Daily Mail* that users of Internet Explorer had lower IQs than people who used other browsers had been the result of an online hoax. The organisation which had originally published the information had only been established the previous month. The media organizations which picked up on the bogus organization's press release had not thought to question its authenticity. The hoax organization posted onto its own website the announcement that it had been set up in July 2011 in order to launch a fake study called 'Intelligent Quotient and Browser Usage.' It added that 'the study took the IT world by storm.' What this hoax demonstrated was the credulity of traditional media organizations in the face of the presumed authority and authenticity of new media platforms.

Another surreal fiction – which originated as a viral email and, somehow legitimized by its online ubiquity, appeared on the websites of various news organizations in January 2005 – concerned the antics of a Slovakian man who reportedly urinated his way to freedom after his beer-laden car was caught in an avalanche. Such stories are immaculate simulacra, copies without originals, relentless reflections which lack material objects. It is precisely because they have no foundation in material truth that it seems suddenly difficult to challenge their claims. Just as media globalization dissolves geographical and cultural boundaries, so a hypermediated society's emphasis upon the virtualized contemporary moment rips narratives from their material history and cannot therefore any longer distinguish between those which have lost that history and those which never had it in the first place. As Wikipedia, Facebook and Twitter have become sites for the generation and dissemination of that which is perceived as information, we have witnessed what Redden and Witschsge (2010: 185) have described as a blurring of the spaces of news and of non-news: a blurring of the material and virtual, of history and fantasy.

In November 2013 it was widely reported across social networking sites that an Indian man had died as a result of being swallowed by a python after falling asleep drunk outside a liquor shop. One particular iteration of the account alone had garnered 15 thousand retweets. For those seeking evidence of this unusual event, the tale was accompanied by a helpful photograph of the immediate aftermath of the incident. However, as *The Huffington Post* pointed out on 29 November, the same insatiable reptile had apparently eaten a Chinaman in August 2012, an Indonesian in January 2013, a South African woman in June and a Malaysian boy the previous month.

In January 2014 a photoshopped image of a 160-foot-long giant squid – which apparently dwarfed its Lilliputian onlookers – was tweeted, retweeted, Facebook-posted, reposted and viewed by at least half a million people within a matter of days. Early 2014 also saw the re-emergence on social media of a story that had been doing the rounds for several years: the myth that a female African-American screenwriter had won cases of copyright infringement against the makers of the *Matrix* and *Terminator* film franchises. In February 2010 *The Examiner* had decried the fact that the story continued 'to be reported on websites as fact.' Indeed as early as July 2005 the *Los Angeles Times* had reported that although the case had been dismissed by a court ruling the previous year there was 'an alternate reality to this story' – one created by chain emails and online news portals – which smacked somewhat of the virtuality of the *Matrix* movies themselves.

In another *Onion*-related case, in November 2012 the Chinese *People's Daily* ran a story headlined 'North Korea's top leader named The Onion's Sexiest Man Alive for 2012' – apparently unaware that *The Onion* is a spoof news site. The Chinese newspaper quoted directly from *The Onion*'s own report in hyperbolic praise of Kim Jong-un: 'With his devastatingly hand-some, round face, his boyish charm, and his strong, sturdy frame, this Pyongyang-bred heart-throb is every woman's dream come true. Blessed with an air of power that masks an unmistakable cute, cuddly side, Kim made this newspaper's editorial board swoon with his impeccable fashion sense, chic short hairstyle, and, of course, that famous smile.'

Yet just as new media technologies may perpetuate misinformation and disinformation, they can also afford the tools to expose and debunk such obfuscations, mystifications and downright lies. North Korea, for example, clearly seeks to deploy these technologies to reinforce its power structures; and yet these technologies are simultaneously deployed to subvert such structures. In an absurd reversal which might have appealed (in different ways) to Mikhail Bakhtin as to Søren Kierkegaard – a sense of licence-as-control is transformed into the possibility of licence-as-liberation and the individual citizen who ridicules power suddenly seems able to overturn the discursive operations of the entire totalitarian state. On 10 December 2010 the BBC reported that on every occasion that North Korean leader Kim Jong-un's name appears on any official North Korean website, it is automatically displayed in a very slightly larger font than the rest of the text. Despite his nation's obsessively tight internet controls, however, the peoples of the Korean peninsula have together developed hybrid high-tech/low-tech methods of subverting these measures. One such tactic has seen

South Koreans attaching USB sticks (containing anything from South Korean TV soap operas to the Korean language version of Wikipedia) to balloons and floating them across the border into North Korea. It has been claimed that the cultural influence of the American television soap *Dallas* contributed to the Romanian people's desire to overthrow the dictatorship of Nicolae Ceaușescu at the end of the 1980s; and in (and out of) the absurdity of North Korean absolutism we might perhaps see the possibility of an equally absurd culture of carnivalesque resistance.

In October 2013 a number of the internet's burgeoning band of citizen-scrutineers claimed that an image of Kim Jong-un and his party bigwigs visiting the site of a children's hospital had been faked – that the people in question had simply been photoshopped into the scene. In March that year online commentators had also pointed out that a North Korean publicity shot of a military exercise – a hovercraft beach landing – appeared also to have been photoshopped – as several of the hovercraft seemed to have been cloned. In January 2011 eagle-eyed web-users had spotted that a PR film for the Chinese air force had included footage from the 1986 film *Top Gun*. On 13 February 2013 *The Independent* reported that critics had expressed doubts in relation to an image of a fighter plane built in Iran on the grounds that it appeared to be a model rather than the real thing: 'the jet could not fly because it was too small and made of plastic.' Iran's powers-that-be responded by publishing an image of the aircraft soaring majestically over a snow-capped mountain peak. But Iran's online cognoscenti remained unconvinced – they were indeed, if possible, even less convinced than before. They pointed out that the two images of the plane were in fact identical – it was just that one of them had, thanks to the increasingly questionable (because increasingly mundane) miracle of photoshop, been superimposed over a picture of an Iranian mountain apparently harvested from the website PickyWallpapers.com.

Nor is the digital exposure of such digital fakery limited to the world's less-than-obviously-liberal regimes. In April 2013, for example, Britain's Conservative Party were embarrassed when it was revealed that an image of 'young Tory activists' featuring on its website in fact comprised a group of Australian students. It seemed that the blatantly healthy and happy demeanours of the young people involved had given the game away.

The new media paradigm is thus in this sense doubly absurd: it is not just that its paradoxes and illusions are absurd in themselves; it is also that its crowd-sourcing structures may promote the exposure of that absurdity. And this leads to a third absurdity: the possibility that weakness might defeat strength, that the individual might overcome the power of the state. And then a fourth and fifth absurdity: that the structures which might afford such resistance also (by giving the illusion that the expressive liberation of the individual has already been achieved through these technologies) reduce the public will for such resistance; and that yet (even despite all this, almost impossibly) these liberating potentials remain and are sometimes (albeit of course tantalisingly rarely) realised.

The geneticist Richard Dawkins (1989) famously observed that ideas (or 'memes') evolve faster than genes because of the speed of their reproduction – the speed, that is, at which they are passed on. The online reproduction of memes clearly further accelerates their reproduction and extends their spread. It therefore increases the speed and variety of mutation, whilst at the same time entrenching (and therefore affording the unpredictable re-emergence) of the most unlikely notions. The most traditional of old media forms – such as the book – are not immune to these processes of transformation and delimitation. In October 2013, for instance, it was revealed (much to the shock of the retail companies involved) that such established booksellers as Barnes & Noble and W. H. Smith had – apparently unwittingly – been selling extreme e-book pornography featuring rape, incest and bestiality through their online sites. Four months earlier it had emerged that a Hertfordshire school librarian had discovered that a poem entitled 'Two sunflowers move into the yellow room' and widely attributed to the late-eighteenth/early-nineteenth century English romantic poet William Blake (and credited to Blake on many school reading lists) had in fact been written in 1981 by the American poet Nancy Willard. The fact that the poem was indeed clearly and explicitly an *homage* to Blake's 1794 poem 'Ah! Sun-flower' makes it seem even more extraordinary that many websites continued a year later to claim that the poem was in fact a Blake original. The librarian in question was quoted by the BBC as saying that this was an instance of how 'the internet can mutate reality.' It is also an example, we might add, of how those mutations – despite all those competing memes that expose them as unworkable fictions – can endure. The internet may, inadvertently or otherwise, spread lies; its citizen-auditors may also come to denounce many of those lies; and yet, once they are out there, despite all those denunciations, those lies tend to survive.

Towards the end of 2012 allegations spread across a number of social media sites that the British Conservative peer Alastair McAlpine was guilty of crimes dating back to the 1980s involving the sexual abuse of children. These allegations were swiftly, clearly and publicly disproven. Lord McAlpine died in January 2014. However, months after his death a Google search on his name may still predict *lord mcalpine twitter*, *lord mcalpine child abuse* and *lord mcalpine paedo*. It is as if his vindication never took place. Like an ingenuous elephant, the internet never forgets; or, rather, it cannot purge its own almost boundless memory of the lies, it forgets that lies are lies, and so those lies appear destined to resurface time and again. The internet has no conscience but it has a memory which is both boundless and irreversible. (Cf. Travis & Arthur 2014.)

The digital village

'Everything,' wrote Baudrillard (1988: 32), 'is destined to reappear as simulation [...] terrorism as fashion and the media, events as television. Things only seem to exist by virtue of this strange destiny. You wonder whether the world itself isn't just here to serve as advertising copy in some other world.' Yet surely we have *always* been mediated, *always* lived in a mediated world? History, society and even human consciousness itself all require networks of communication – languages, media – in order to exist. So what makes the situation so different today?

What makes all the difference, a world of difference, is the fact that today's globalized multimedia are coming to converge into a single monolithic *omnimedium*. Stuart Hall (1980) wrote famously of the possibilities of negotiated and oppositional readings of media texts; but these active readings are, of course, only possible if we are exposed to a range of discourses, a variety of meanings. So what happens when we have become reliant upon – and indeed inscribed within – a single medium, an all-encompassing code? Like Jim Carrey in *The Truman Show* (1998), how can we see things clearly when we are part of the picture, when the medium is our entire world?

When examining the impact upon society of new paradigms, structures and technologies, there is perhaps no place better to look than speculative fiction. Popular science fiction has, for more than a century, explored the unfolding future of western civilization, and its contemporary incarnations continue to investigate these problems and possibilities. Now more than ever, these fantasies articulate the absurdity of emergent paradigms of reality. In, for example, the opening episode of the American science fiction TV series *Caprica* (2009) – a prequel to the reimagined *Battlestar Galactica* (2004–2009) – a young woman on a world wrought by ethnic tensions is killed in a terrorist attack perpetrated by a monotheistic, fundamentalist and extremist religious cult. Her father's attempts to memorialize her – to concretize and externalize her memory – at first through her reconstruction in a virtual world, and then by transferring that reconstruction into the physical world, set off a chain of events which leads towards the utter transformation and then the near extinction of human civilization.

In this context, one particularly apposite (although not particularly successful) television series was the 2009 reimagining of the 1960s Cold War fantasy *The Prisoner*. In the remake a New Yorker – an electronic surveillance expert – discovers himself in a surreal new world, a village at the heart of a desert, a community of false bonhomie as superficial and inescapable as Facebook, a global village closer to Baudrillard's desert of the real than to McLuhan's utopian vision of informational mobility and interactivity – Abu Ghraib meets Disneyland via *Lost*. This character was played by the American actor Jim Caviezel – who would go on to star in the series *Person of Interest* (2011–), in which he would also play an intelligence expert caught up in a fantastically sinister world of unrelenting and omniscient electronic surveillance.

This post-historical protagonist cannot access the past; his memories are fragmentary and incoherent. The last thing he can remember is an explosion in New York. After the devastation of 9/11 only the absurdity of this hypermediated reality and its simulacrum of freedom remain: 'there is no New York,' says the leader of this new world: 'There is only the village.'

The paradigm has irrevocably shifted, and this echoes a sentiment witnessed in so many post-9/11 screen fantasies – from *Battlestar Galactica* and *Star Trek: Enterprise*, through J. J. Abrams's *Cloverfield, Lost, Fringe, Believe, Almost Human, Person of Interest* and *Star Trek* films, to *Heroes, Jericho, Revolution* and Christopher Nolan's Batman trilogy – a sentiment epitomized in 2013's pilot episode of *Marvel's Agents of S. H. I. E. L. D.*: 'The battle for New York was the end of the world. This is now the new world.' Or, as Tony Blair had put it in October 2001, 'the kaleidoscope has been shaken, the pieces are in flux, soon they will settle again. Before they do, let us re-order this world around us.' A decade earlier the first President Bush had announced his dream of a 'new world order' – and twelve years before that, as he declared his candidacy for the Republican presidential nomination, Ronald Reagan had famously quoted Thomas Paine: 'We have it in our power to begin the world over again.'

The Prisoner's village, like the Americentric global village envisaged by these neoconservative dreams of a new world order, suffers a permanent condition of heightened paranoia in which everyone lives in fear and in which such fear is publicly perceived as 'guilt in disguise'. Everybody distrusts everybody else, and everyone spies on each other, even parents on their children – and vice versa. This surveillance society founded upon a self-sustaining state of terror offers overt contemporary parallels: the West has reinvented the strategies of its abstract enemy in order to apply them not only to that imaginary other but also to its own populations. In *The Prisoner*'s village, names have been replaced with numbers: humanity literally has been digitalised – just as contemporary history sometimes feel as if it has been converted into a sequence of key numbers: 9/11, 7/7, 2012 (for some the London Olympics, for others an averted apocalypse – for many, both).

The series's opening episode ends with a vision of the twin towers in the heart of the desert, a mirage of a lost world, a faded paradigm which can no longer be reached. In the programme's fourth episode cracks begin to form in the reality of the village – actual cracks, holes in the ground on which the village is built. Beneath the surface there is absolutely nothing, a void. This truly is Baudrillard's depthless desert of the real. The village is, as Jean Baudrillard (citing Borges) might have had it, a full-scale map of the real designed to map directly over the real until the real recedes and the map becomes real, the only real at least which remains, and eventually (as history diminishes) the only reality that there has ever been (Baudrillard 1994: 1). It is a virtual realm, which, like the electronic state, the digital game, reality television and the social networking site, reinforces and exacerbates the very problems and paradoxes it had purported to resolve. The village is, as such, a map of the multimedia mapping of the world, a model of the internet itself, or of the world for which the internet has become a metaphor and which at the same time is merely a metaphor for that network.

The primacy of this virtual map and its potential for disparity with the physical world became jarringly clear when in June 2010 a Californian woman called Lauren Rosenberg launched litigation against Google after having been hit by a car on a four-lane highway while using a pedestrian-friendly route recommended by Google Maps. According to her lawyer, Google had 'created a trap with walking instructions that people rely on. She relied on it and thought she should cross the street.' By doing so Ms Rosenberg fell victim to the distance and tension which still persists between the material world and its map – or rather, insofar as the map had become her dominant paradigm, between the map and its material reflection.

In November 2012 a team of Australian scientists discovered that a sizeable South Pacific landmass known as Sandy Island – which featured on Google Earth and Google Maps – did not in fact exist. They seemed quite perplexed by revelation. Expedition member Maria Seton told the BBC: 'It's on Google Earth and other maps so we went to check and there was no island. We're really puzzled. It's quite bizarre.'

A year later – in November 2013 – fans of the BBC television series *Doctor Who* were able (via Google Street View, and in celebration of that programme's fiftieth anniversary) to enter a police box parked on London's Earl's Court Road and to find the greater dimensions of a futuristic alien time machine held miraculously within. This served as a useful reminder that the world according to Google – the world many have come to trust as their definitive version of reality – can so easily be subverted into an advertent or inadvertent fantasy.

Yet Google's virtual worlds may sometimes seem not merely too fantastical but also sometimes far too real. Also that month (in November 2013) Google Earth agreed to remove a satellite image of the body of a teenager murdered in Richmond, California, which had appeared as part of its photographic catalogue of the planet. Excessive material reality is not what their users expect or desire from such virtual services; and so we reject anything that does not fit into this controlled artificial world.

That same month Andre Wisdom, the captain of England's under-21 soccer team, had failed to live up to the promise of his patronym when he had been obliged to abandon his £100,000 Porsche as a result of his over-reliance upon his car's satellite navigation system to chart a route to his own club's football stadium. The *Daily Mirror* reported that the 20-year-old had 'got stuck in a huge mud-filled puddle after driving down a woodland track three miles from the nearest main road.' The following month a Taiwanese tourist had accidentally walked off a pier in Melbourne and fallen into Port Phillip Bay while checking her Facebook page.

James Boswell (1986: 273) once gave an account of his friend Dr Johnson's attempt to refute Bishop George Berkeley's 'ingenious sophistry to prove the non-existence of matter, and that everything in the universe is merely ideal.' Johnson's strategy was to kick with all his strength against a large rock, and as his foot rebounded (no doubt mighty painfully) therefrom, he declared: 'I refute it thus.' It is with a similarly steadfast stubbornness that today we will be hit by cars, drive into bogs and fall into the sea in order to prove the opposite truth: that in fact our reality is merely a virtual ideal. And as we await the arrival of the ambulance, tow truck or lifeboat, we can rest content and secure in the certainty that – however painful and messy and dangerous the experience may seem – the virtual world is somehow safer, surer and apparently within our control than material reality ever might be.

When, therefore, we are reminded of the limitations and failings of such technologies, the shock that this recognition incurs seems even in the most extreme cases to outstrip our horror at the human tragedy involved. The disappearance, then, of Malaysia Airlines flight MH370 in March 2014 highlighted the inadequacies of contemporary information and communication technologies, raising urgent questions (as the BBC put it on 17 March) as to 'how a modern aircraft packed with communications equipment can apparently vanish without a trace.' Further to this, when on 24 March Malaysia Airlines sent a text to the families of those lost announcing that 'none of those on board survived' serious questions were also raised as to the appropriateness of this virtual communication strategy: as Fox News reported the following day, the airline's CEO had been obliged to defend this approach in response to an outcry of media criticism.

By contrast, when, the following month, a South Korean ferry sank, killing many of those on board, it was widely reported that children aboard the stricken vessel had texted final 'heart-breaking' messages to their parents as the ship went down: 'this might be my last chance to tell you I love you.' It is difficult to reconcile our diametrically opposed responses to these contextually related (but morally distinct) uses of the same piece of mobile communications technology. Our attitudes towards such technologies are clearly both ambivalent and evolving.

The era of the satnav and the SMS does not, after all, hold all the answers. These technologies maintain the capacity to lose us – they cannot save us – they are not 100 per cent reliable – not only because they do not always work, but also because they do not provide the solutions to all our problems even on those occasions on which they are functioning properly in themselves. The real problem is that we of course sometimes forget this: that we rely entirely on such technologies to get us where we want to go – whether that is a utopian societal future or a fulfilled social life or just down the road to the shops. (Cf. Markoff 2014.)

The problem, some might argue, is not then the map, the satnav, the SNS or the SMS or any other virtual tool: it is the fact that the rest of the physical world has not as yet fully recognized the primacy of the virtual realm's authority. The 2009 TV series *The Prisoner* similarly envisioned a realm in which that authority would go almost entirely unchallenged, an imaginary landscape whose absurdity only one individual – the series' protagonist – appears to recognize.

The Prisoner's unreal village is a shallow dream, the simulacrum of, as its custodian explains, a 'new world' invented to replace the pain and complexity of material history – because 'the world is not a pretty thing when you look at it too close.' *The Prisoner* imagines an escape from contemporary reality, and, like the *Matrix* films, it addresses at once the unsustainability of the illusion and, conversely, the threat that the illusion may become overwhelmingly real, may become the only reality left to us. The final horror of *The Prisoner*, its final revelation, is that in its closing episode the series's protagonist, the village's lone voice of resistance, the eponymous prisoner himself, swaps sides to become the gaoler, the born-again zealot and custodian of its pseudo-utopian illusion.

Fantasy has traditionally offered itself as an allegory of, or as a satire upon, urgent contemporary concerns; but it may be that, at the extremes of history (when history reaches its own extremities at its borders with the fantastical, the post-material and the virtual) the fantasy space itself becomes almost indistinguishable from the historical. As the extraordinary renditions of fantasy approach those of fact, so fact becomes ever more fantastical.

Science fiction then is here. The long-awaited retail sales launch of Google Glass on 15 April 2014 was accompanied by an announcement that the company also had a patent pending on a contact lens version of the technology – and that, like something out of *Star Trek: The Next Generation*, this could eventually be used to provide information on the visual world to the blind. In August 2013 *The Daily Telegraph* had reported that a New York company had developed a Jewish-themed app for Google Glass: 'JewGlass reminds users of Shabbat start and end times, gives walking or driving directions to nearby synagogues and brings up prayers on screen if no prayer book is to hand.' In February 2014 it was announced that Virgin Atlantic staff were using Google Glass in a 'pilot scheme' – but that this scheme did not extend to the airline's pilots. The same month it was reported that the New York Police Department were trialling Google Glass in support of law enforcement activities. This was also the same month which saw the release of José Padilha's remake of the 1987 science fiction film *RoboCop*. Such fantasies no longer seem particularly fantastical or, for that matter, prescient. And they barely seem significantly more absurd than the world which has spawned them.

CHAPTER 2

Electronic Politics

The French sociologist Pierre Bourdieu (1991: 172) has argued that a lack of access among the general populace to the tools necessary for political participation has resulted in the concentration of political power as the province of a small elite. Although much has been claimed for the potential of new media technologies to promote democratic political participation, it remains unclear whether the application of these technologies in practices as apparently diverse as those of electronic government, interactive entertainment and virtual socialization indeed offer the popular dissemination of the technological and cultural capital which Bourdieu sees as essential to the processes of democratization – or whether they in essence divert their subjects from such processes.

Bourdieu (2005: 62) has proposed that 'to make a decisive contribution to the construction of a genuine democracy [...] one needs to work towards creating the social conditions for the establishment of a mode of fabrication of the general will [...] that is genuinely collective [...] based upon the regulated exchanges of a *dialectical confrontation* [...] capable of transforming the contents communicated as well as those who communicate.' It appears that the homogenizing seamlessness of contemporary media technologies refutes the possibility of any such dialectical confrontation. Those who might see the potential of emergent information and communications technologies to foster a global village in which these technologies unify society's fragments (McLuhan 2001: 385) – or for that matter a return to the 'vibrant democratic intellectual culture of the eighteenth-century London coffeehouse' (Keen 2008: 79; see also Doyle and Fraser 2010: 229) – might note that the possibilities of a Habermasian public sphere – 'the sphere of private people come together as a public' (Habermas 1989: 27) – seem to have receded: 'the communicative network of a public made up of rationally debating private citizens has collapsed; the public opinion once emergent from it has partly decomposed into the informal opinions of private citizens without a public and partly become concentrated into the formal opinions of publicistically effective institutions' (Habermas 1989: 247). The internet does not appear, for all its claims, to have reversed this trend.

These new technologies do not in themselves afford a return to a determining sphere of public debate; on the contrary, they appear to offer an illusion of individual public agency (an agency in fact held by institutional bodies) which consigns the individual to an extended and ubiquitous 'private sphere' – as Zizi Papacharissi (2010) has called it. Indeed, Papacharissi (2010: 167) notes that while the concept of the private sphere may offer to describe the processes of civic interaction in post-industrialised societies one should not thereby mistake it, in itself, for 'a recipe for democracy.'

This view does not appear entirely to have reached the political classes. In June 2009 Britain's then Prime Minister Gordon Brown published an article in *The Times* newspaper in which he argued that internet access for the entire population of the UK was an essential factor in a bid to secure the healthy economic and democratic future of the nation. Brown wrote that 'a fast internet connection is now seen by most of the public as an essential service.' He added that 'digital Britain cannot be a two-tier Britain – with those who can take full advantage of being online and those who can't.'

In July 2010 Brown's successor David Cameron backed the Manifesto for a Networked Nation, a report produced by the government's Digital Champion Martha Lane Fox which announced the need to get all British people of working age online by 2012 (a goal which was not reached): 'digital inclusion [Cameron stressed] is essential for a modern dynamic economy.' Neither Brown nor Cameron went so far as France's Constitutional Council which had, in June 2009, ruled that internet access – as 'an essential tool for the liberty of communication and expression' – represented a fundamental human right. Both, however, shared the notion that access to the internet must be available to all in order to maintain conditions for the ongoing development of a sustainable and equitable modern state.

That comprehensive access still seems a long way off. The UK's Office for National Statistics reported that 83 per cent of British households had internet access in 2013, with just over half of those (only 42 per cent) having access to a broadband connection. Those figures also showed that six per cent of people aged 45–54, 11 per cent of people aged 55–64 and 36 per cent of those aged 65 and over never used the internet at all. In August 2012 it had been reported that a number of candidates in UK elections for police commissioners were calling for public funds to support a mailshot to all voters on the grounds that it was 'undemocratic' to rely entirely on the online provision of information when – according to a report from the Electoral Commission earlier that year – the seven million voters without internet access could be denied relevant information. Internet access in the UK is still far from universal, and seems divided along socio-geographic lines. In September 2013 a report from the charity Age UK demonstrated the existence of a north-south divide in the number of pensioners who use the internet. The report showed, for example, that over-65s in Surrey were more than twice as likely to have internet access than those in Tyne and Wear. The UK also suffers an urban-rural divide in the provision of internet access: as the BBC's technology reporter Jane Wakefield observed on 26 September 2013, 'making sure that those living in the countryside get broadband speeds comparable to those living in towns and cities has long been something the government has grappled with.'

Might digital exclusion and imbalance therefore undermine the prospects for democracy in a contemporary society? In their study of the uses by members of the U. S. Congress of social networking technologies for establishing dialogues with their constituents, Glassman et al. (2010: 11) note that such online dialogues might become influential in the determination of legislative voting and policy decisions. In his study of similar systems in the British Parliament, Williamson (2009: iii) however observes that while new technologies can improve communication flows with constituents it remains the case that digital media do not represent an unproblematic panacea. Clearly one of the ongoing problems with these tools is that they remain in the hands of the computer-literate and internet-active section of any electorate; and that it is conversely those constituents who lack this intellectual and technological capital whose situations may most urgently need to be considered in the development of policy agendas.

This is not, for example, to disparage the hopes of such theorists as Coleman and Blumler (2009): aspirations that new media technologies may eventually play a part in restoring a disillusioned public's trust in the structures and institutions of democratic politics; but it is to emphasize that these technologies cannot achieve such a goal in themselves – and that, if we assume they can, opposite tendencies may emerge.

The new democracy

A former senior engineer at the website, Karel Baloun (2006: 88) has asked whether Facebook will 'somehow change the world, starting with its political election-focused groups and candidate profiles.' What is perhaps most interesting about this question is the fact that it is being asked: that fact that we might seriously be expected to take the political impact of such a social networking site quite so seriously.

In April 2010, four weeks before a British parliamentary election, it was reported that the UK's Electoral Commission had collaborated with Facebook to encourage unregistered voters to enfranchise themselves. Facebook users were asked if they had registered to vote, and if they had not, were directed to a page that would allow them to register online.

Can the use of such social networking sites as Facebook promote participation in democratic processes? And should it do so: should we be specifically targeting those with internet access in voter registration? On the face of things, such high profile political endeavours as Barack Obama's electioneering pages on Facebook would seem to suggest that such sites can foster democratic activity. There remain, however, significant questions as to whether the fundamental ideological perspectives underlying and advanced by these homogeneous media forms are themselves consistent with western notions of democracy, diversity and socio-political engagement.

On 15 April 2010, immediately after the UK's first ever televised election debate featuring the leaders of the country's three main political parties, BBC Technology Correspondent Rory Cellan-Jones reported via

Twitter the news that 36,483 people had posted a total of 184,396 messages onto that platform in relation to (and during) that debate. In an article on the *BBC News* website the previous day Cellan-Jones had pointed out that Facebook would meanwhile run its own digital version of the leaders' debates – and that all three party leaders had 'agreed to answer questions submitted by users.' The morning after the first debate, on the BBC's *Breakfast* news programme, Cellan-Jones added that Facebook had experienced a server overload as a result of online activity related to the debate. Yet Cellan-Jones also suggested that this activity did not represent a step change in political participation – it had merely shifted extant offline debates onto these sites. One might, however, ask whether the limitations and structures of these sites allow for the range, extent and freedom of debate possible offline – or whether they offer a regulated imitation of and replacement for such participation practised in the non-virtual public sphere. Answering questions through Facebook (or indeed allowing one's advisers to do so) seems rather less onerous a task for a politician than facing the public and their political opponents in a live television debate – or facing the opposition in parliament – or facing journalists in interviews or press conferences.

Guardian journalist Peter Bradshaw commented to the BBC in April 2010 that the UK election campaign (then ongoing) had witnessed the sudden collapse of the traditional press power to set the news agenda: 'these reality-TV-style debates have overnight completely changed the political landscape.' The point is not simply that these debates moved political agenda-setting into the realm of television (television had shared this role with the press for some decades), but specifically that this process had been appropriated by an interactive multimedia format which owed much to the style and structure of reality television. Furthermore, these televised election debates had at once blurred the distinctions between, and in doing so drawn attention to the tensions between, older and newer media paradigms: between the conflicting etiquettes of broadcast television and narrowcast online discourse. In the immediate wake of the first of these live leaders' debates, it was, for example, reported that ITV news had accidentally broadcast an obscene message abusing Conservative Party leader David Cameron: a shot of the readers' discussion page on the station's debate webpage which focused on a particular post which read: 'Cameron – You're a first class cunt.'

The potential entertainment value of the interactive multimedia circus which had grown out of this democratic process did not go unnoticed by commercial interests. During Britain's general election campaign of April–May 2010, the company producing the yeast-based spread Marmite created parallel Facebook pages for the Marmite Love Party and the Marmite Hate Party, complete with slogans ('Spread the Love' and 'Stop the Spread' respectively) and campaign statements, manifestos, pledges and profiles of the party leaders – as well as links to the Marmite News Network which kept readers updated on the latest campaign news. Users with strong positive or negative opinions on the product could log onto Facebook to register their vote for or against the spread. By the end of the election period more than 150,000 people had registered their support of the Marmite Hate Party's campaign – while the Marmite Love Party had attracted over 350,000 supporters. Is this then the extent of political engagement offered by Facebook – a pastiche of engagement which at once trivializes and sublimates the desire for socio-political agency?

In April 2010 it was reported that the manufacturers of Marmite were threatening to take legal action against the extremist British National Party in order to ban its use of an image of a jar of the yeasty spread in a political broadcast. Marmite's spoof campaign had itself been hijacked by the anti-democratic extreme of British political opinion; the parody of democratic processes had become a tool for forces actively seeking to undermine those processes. This seems an appropriately absurd model for the effects of hypermediation upon the processes of politics.

Everyone who is anyone in public life has to have a Facebook page. The British royal family launched theirs in November 2010, shortly after George W. Bush had launched his own, but many public figures (from Barack Obama to Gordon Brown) have tried (with varying degrees of success) to exploit the site for rather longer. By November 2010, six months after his election to office, British Prime Minister David Cameron, for example, had over 92 thousand fans of his Facebook page. It may be noted however that at the same time the Facebook page entitled 'throwing eggs at David Cameron, brick shaped eggs – made from brick' had over 99 thousand fans, while 'David Cameron is the new PM. Last one out of the country turn off the lights' boasted over 108 thousand fans and more than 204 thousand people supported the page entitled 'David Cameron wants change! Give him 30p and tell him to f**k off.' Although David Cameron's own Facebook page had, four years after his election, accumulated just over 180 thousand likes, it was reported in March 2014 that the Conservative Party had spent thousands of pounds on Facebook ads in an attempt to increase the numerical support for Cameron's page.

One might argue that, if political participation through Facebook or other online activities is meaningful and empowering, then such activity biases democratic processes in favour of those groups in society (the economically, technologically and educationally advantaged) who least need that empowerment. Conversely one might assert that, if such participation merely offers an illusion of empowerment and socio-political agency which trivializes and denigrates politics itself, then it in fact undermines the desire, and therefore the potential for, real empowerment and meaningful agency. The internet user feels as though she is at the centre of the universe, and this imaginary repositioning pre-empts the subject's struggle towards such centricity and significance.

In the immediate wake of the British general election of 6 May 2010 – which resulted in a hung parliament and emergency talks between the Conservative Party and the Liberal Democrats over the possibility of a coalition which would make the formation of a government possible – a Facebook group entitled 'We don't want the Liberal Democrats to make a deal with the Conservatives' was established. According to the founders of the group, 'the Lib Dem head office advised us to start a group to get an idea of how many people object to the proposed coalition.' After five days this group had over 60,000 members – a figure dwarfed by the support shown to Marmite, but still perhaps a not entirely insignificant representation of the electorate (nearly one per cent of those who voted for the Liberal Democrats at the election). A similar Facebook group named 'We don't want the Liberal Democrats to make a deal with Labour' was also started: within the same time period it had attracted a total of 83 members. In addition to this a Facebook page called 'We don't want the Liberal Democrats to make a deal with the Labour Party' had attracted 203 supporters and another page called 'We don't want the Liberal Democrats to make a deal with Labour' had gained 59 supporters. It therefore appeared that those against a pact with the Conservatives very heavily outnumbered those opposing an agreement with Labour. At the end of the fifth day after the election, however, the Liberal Democrats abandoned any possible agreement with the Labour Party when they announced a deal with the Conservatives. So much then for the power of Facebook: so much perhaps for the internet's capacity to channel public opinion directly to the political elite – so much, at least, for that elite's willingness to take that opinion on board.

Michael (2013: 46) has supposed that such social media platforms as Twitter might one day come to 'support the development of a more collaborative political culture' – but that 'any such process would require authenticity on the part of politicians, informed contributions from the public, and a willingness to engage from both.' In their analysis of non-campaigning tweets by members of the UK parliament, Knowles and Glennon (2014) have found that policy-related posts comprised a 'disappointingly low' proportion of only 7.4 per cent of their overall sample, but that politicians engaged instead with 'lots of locally focused tweeting about sport as a way of connecting with the local community.' This chimes with the findings of research by Margaretten (2014) which has suggested that rather than using the platform to foster policy debate MPs have tended to deploy Twitter as a promotional tool to bolster a sense of their own normative public authenticity – to be seen as people of the people. Knowles and Glennon have stressed that these political uses of Twitter are only 'symbolically interactive' – adding that 'what they are not is dialogic: the proportion of dialogic tweeting – that kind of interactivity – is very low.' In April 2014 the Hansard Society's eleventh annual Audit of Political Engagement reported that, while 67 per cent of people surveyed thought that politicians were out of touch with the lives of ordinary people, 44 per cent would welcome the introduction of open Q&A meetings (in person) between parliamentarians and their constituents, but only seven per cent called for uses of social media to address the disconnections between MPs and the electorate.

If the politicians' microblogs do not look set to rekindle the flames of public policy debate, then perhaps the weblogs of online political commentators might ignite some sparks of democracy. The well-known British blogger Iain Dale has suggested that the political blog has promoted levels of political engagement among younger voters in the UK: 'I think blogs have engaged people who might otherwise not have become involved in the political process, especially among the 18–25 age group. I know this from dozens of emails I receive from young people, who may be turned off engagement with political parties, but want to involve themselves in the democratic process. They see blogs as a medium for doing this.' Although Dale does not see the blog replacing or even seriously rivalling traditional news services, he acknowledges the weblog's increasing role in setting the news agenda: 'I don't believe that blogs will replace old media, but there's no doubt that new media has had a major impact on the old media. They have provided the old media with new voices to report and engage with and use as independent pundits and commentators.' Dale also recognizes that the weblog phenomenon does not necessarily represent an alternative media form which promises the democratic redistribution of political influence away from established voices: 'In some ways mainstream journalists have already colonised the blogosphere so blogs have, to an extent, already become institutionalised. This is inevitable and is neither a good nor a bad thing. It's just a fact.' Modes of resistance to institutionalised power themselves become institutionalised and entrenched: indeed, Dale himself is now something of a recognized figure of the media establishment.

The rise of the independent blogosphere has corresponded with a certain amount of media hype surrounding a phenomenon which has become known as citizen journalism. The excitement about citizen journalism ignores the fact that that it does not really exist – not at least as a replacement for professional news organizations. The successful citizen journalist tends to become a professional journalist; the successful independent online newssite tends to become increasingly professional (the *Huffington Post* was, for example, bought by AOL in 2011 and won its first Pulitzer Prize in 2012). What has been hailed with such hyperbolic enthusiasm as citizen journalism often constitutes not much more than uninformed and little read opinion pieces which languish obscurely on individuals' personal blogs, or the capturing of photographic images or video footage on mobile phones (and sometimes the posting of this raw material onto public websites) which is then used (as witness testimony has always been used) by professional journalists.

While this latter development has, as Silverman (2014) observes, proved a useful resource in the news-gathering process, it hardly represents a culture shift in the way that the majority of people receive their daily news. Few news-consumers directly trawl Blogger, Flickr, Twitter or YouTube for raw news. This material is usually accessed from the services of traditional journalistic outlets, after it has undergone the time-honoured processes of professional mediation (selection, editing and interpretation) – and this may then be accessed by news-consumers from these institutional sites, albeit increasingly via recommendations posted on such sites as Twitter and Facebook. Approximately 20 per cent of online news consumers arrive at stories via social media (and this figure can double if a story goes viral), and it should therefore come as no surprise that shrewd news institutions are increasingly collaborating rather than competing with such portals (Ashton 2012). Thus the public may be at the gates, but the gates are still being kept by the traditional authorities.

Digital delivery

The availability of the electronic provision of commercial services (in, for example, the financial industry or the information supply sector) is virtually ubiquitous in the post-industrial world, and the replication of this model in the public sector (in the form, for example, of online tax returns, direct debit digital payments for state services and the electronic dissemination of public service information) is similarly common. As Coleman (2005c: 207) suggests, as commercial activities have been radically reshaped by these new technologies, it would have been unlikely that processes of political representation would not also be influenced by these trends. Thus the commercial e-service model has been translated into the area of e-government. The successful adoption of this paradigm by political parties (such as by Barack Obama on his journey to the White House) is one manifestation of this phenomenon. Barack Obama's first presidential campaign has been described by Papacharissi (2010: 1) as having achieved a degree of 'media-savvy notoriety' for its innovative deployment of new media technologies. The Obama model involved not only online fund-raising and campaigning activities, but also the possibility (or at least the semblance) of participation in policy debates (on, for instance, the candidate's Facebook page) – although it is uncertain to what extent (if at all) these online debates might have framed the presidential agenda. In February 2012 *The Guardian* newspaper reported that Obama's campaign for re-election was deploying Facebook as a key strategic tool: the website offered the campaign 'a source of invaluable data on voters' in order for it to develop 'a vast digital data operation that for the first time combines a unified database on millions of Americans with the power of Facebook to target individual voters to a degree never achieved before.' But Obama's love affair with Facebook was not set to last forever. In January 2014 Barack Obama was quoted as suggesting that young people 'don't use Facebook any more' and that March – following revelations that the National Security Agency had been harvesting data from the site, and following a high-profile meeting between President Obama and Mark Zuckerberg to discuss the issue – the company had described the U. S. government's reforms to its online surveillance policies as 'simply not enough.' The potential then for such sites to allow political elites to connect with their electorates is both unpredictable and open to abuse.

Proponents of electronic politics and electronic government have argued for their potential to move beyond the commercial sales-and-supply model towards a dialogical paradigm. E-government's advocates announce the possibility of something more than e-services; they suppose that it (not unattractively) offers the chance of e-democracy – of interactive popular participation in political processes: 'interactivity opens up unprecedented opportunities for more inclusive engagement in the deliberation of policy issues' (Coleman 2005c: 209). This is a vision in which, as Coleman and Spiller (2003: 11–12) propose, 'interactive technologies, which can facilitate online consultation and dialogue, make direct representation a possibility.'

But this is only the beginning. If democracy's ultimate expression and test take place at the ballot box, then electronic democracy may be seen as coming of age in the introduction of online voting. Stephen Coleman (2005b: 97) writes that, while online voting may be take place in traditional polling stations, it can also be employed, more ambitiously and radically, as a means to afford voters the opportunity to participate in elections from home or at work or wherever they can access the internet.

Risks of course accrue to this process. Coleman (2005b: 96) pertinently enquires as to the nature of these risks: 'are risks purely technological or are there other processes at stake?' As he goes on to suggest, these risks (which technologists may see primarily in terms of systems security) may become matters of *moral* security – when for example, voters without internet access might consider the limitations upon opportunities for online voting as representing 'an act of discrimination' (Coleman 2005b: 99).

Some argue that those moral risks are outweighed by the possible benefits for whose sake those risks might be incurred. Those perceived or projected advantages tend to based upon ambitions of increased civil and democratic participation prompted by the uses of these technologies: that these technologies might offer 'a counterbalance against voter apathy and therefore increase voter turnout, which in turn legitimises the outcome of the electoral process' (Xenakis and Macintosh 2007: 14).

However, as Xenakis and Macintosh go on to stress, this hypothesis remains unproven. In, for example, the local elections of August 2007 the British Electoral Commission ran a series of pilot schemes involving online and telephone voting services in the districts of Rushmore, Sheffield, Shrewsbury and Atcham, South Buckinghamshire and Swindon. The Electoral Commission's subsequent reports agreed that the schemes had 'a negligible impact on turnout.'

Writing in *The Guardian*, however, on 26 March 2014 head of the Electoral Commission Jenny Watson suggested that the introduction of internet voting in the UK might be a way a countering the low voter turnout amongst the under-25s – whilst at the same time warning that we should not come to equate 'the act of voting in an election with shopping online – or indeed with voting in *The X Factor*.' The increasingly blurred relationship between democracy and reality television is one to which we shall return.

The Reinforcement of power

In their study of *Computer Ethics and Professional Responsibility*, Bynum and Rogerson (2004b: 318) address the crucial question as to whether the internet will sponsor global democracy or become a tool for the manipulation of the general public by political elites. Bynum and Rogerson (2004a: 6) describe a contemporary scenario in which academic opinion on these questions is polarized between cyberoptimism and cyberpessimism:

> Optimists point out that information technology, appropriately used, can enable better citizen participation in democratic processes, can make government more open and accountable, can provide easy citizen access to government information, reports, services, plans, and proposed legislation. Pessimists, on the other hand, worry that government officials who are regularly bombarded with emails from angry voters might easily be swayed by short-term swings in public mood [...] that dictatorial governments might find ways to use computer technology to control and intimidate the population more effectively than ever before.

In April 2012, for example, *The Independent* newspaper ran as it front page story the news that the British government were introducing legislation to afford police and security services the power 'to watch you on the web' – 'to check on citizens using Facebook, Twitter, online gaming forums and the video-chat service Skype.' The paper took some relish in retrospectively quoting David Cameron – a year before he had become Prime Minister – railing against the UK's surveillance society and arguing that 'we are in danger of living in a control state.' The following day, writing in *The Sun* newspaper, Home Secretary Theresa May attempted to reassure the British public that 'no one is going to be looking through ordinary people's emails of Facebook posts' – only those of serious criminal suspects. One is tempted to ask how Ms May was planning define those 'ordinary people' in the first place. The same day *The Sun* featured comments from another senior Conservative politician warning that this development might turn Britain into 'a nation of suspects' – and from the government's Information Commissioner suggesting that this move represented a 'step change in the relationship between the citizen and the state.'

In December 2012 the Deputy Prime Minister Nick Clegg said that these plans required a 'fundamental rethink' in order to ensure 'the balance between security and liberty.' In May 2013 it was reported that the world's five biggest internet companies (including Google and Facebook) had informed Ms May that they would not be willing to co-operate with her 'snooper's charter'.

That same month the former CIA and NSA operative Edward Snowden leaked to journalists the details of massive global electronic surveillance operations conducted by the American government in collaboration with European security services and international communications organizations. In June 2013 the BBC reported that Google had 'issued a strong denial that it allows the U. S. government to access its servers' and that 'Facebook and Google insist they did not know of PRISM surveillance program' – a mass data-mining operation launched by the NSA in 2007 and operated in cooperation with Britain's GCHQ. On 6 September 2013 the BBC added that, again according to Snowden's disclosures, 'U. S. and UK intelligence have reportedly cracked technology used to encrypt internet services such as online banking, medical records and email.' It added that 'the encryption techniques targeted are used by popular internet services such as Google, Facebook and Yahoo' and 'in 4G smartphones, email, online shopping and remote business communication networks.' In November 2013 the founder of the world wide web Sir Tim Berners-Lee called for a 'full and frank public debate' over these internet surveillance activities: 'any democratic country has to take the high road; it has to live by its principles.' It should therefore come as little surprise that in the run-up to the first retail sales of Google Glass to the general public *The Daily Telegraph* (on 11 April 2014) observed that 'while many enthusiasts see Glass as the future of technology, there are also those who view the device as a spine-tingling enabler of mass surveillance.' Indeed, one might suppose that there is no contradiction inherent to these two positions: that the fear may be that mass electronic surveillance does indeed represent the future of technology. As Slavoj Žižek (2013) pointed out, Edward Snowden's revelations merely confirmed our deeply held suspicions. That was precisely why they were so shocking, because they uncannily exposed what we already knew: 'it is a little like knowing that one's sexual partner is playing around – one can accept the abstract knowledge, but pain arises when one gets the steamy details, pictures of what they were doing.'

Google's Eric Schmidt and Jared Cohen wrote in *The Wall Street Journal* on 19 April 2013 that 'dictators and autocrats in the years to come will attempt to build all-encompassing surveillance states, and they will have unprecedented technologies with which to do so. But they can never succeed completely. Dissidents will build tunnels out and bridges across. Citizens will have more ways to fight back than ever before.' But we might reasonably ask whether the interests and activities of such unaccountable transnational megacorporations as Google, Facebook or Twitter offer ordinary people tunnels or just cunningly disguised bars. Do their fine words do much more than gild the cage?

On 1 February 2014 John Naughton, writing in *The Observer* on the occasion of the website's tenth birthday, proposed that Facebook's business model was analogous to that of the NSA: 'Both need to use surveillance of both intimate and public online activity to make inferences about behaviour. The NSA claims that this enables it to spot and thwart terrorism and other bad stuff. Facebook's implicit – but rarely explicitly articulated – claim is that intensive monitoring of what its users do enables it to both tailor services to their needs and provide precise targeting information for advertisers.' Naughton noted that while the efficacy of the NSA's strategies might be questionable, the profitability of Facebook's plan was evident: during 2013 the company had secured revenues of $7.87 billion. In the first quarter of 2014 alone Facebook revenues reached $2.5 billion.

What then might governments learn from this extraordinary success?

Online government

The risks of electronic government may often appear clearer than the benefits. The political scientist John Curtice (2009) has pointed out that 'people's uses of the internet are primarily a function of their prior motivations. The internet isn't bringing into political activity people who aren't engaging in it offline.' Papacharissi (2010: 105) has argued that 'digitally enabled civic activity has not been associated with an increase in political participation [...] nor has it been identified as a factor in reducing voter cynicism and apathy.' Gibson et al. (2004: 8) suggest that it is highly questionable whether e-government has any effect on democratic participation. As Nixon (2007: 29) points out, research studies have suggested that between 60 and 85 per cent of e-government projects can be regarded as failures. More problematic, however, than the mere failures of new technologies to foster civil society and democratic participation are suggestions that these systems may in fact undermine opportunities for equitable popular representation.

As well as increasing public access to information, new media technologies may, for example, allow governments and economic interests greater powers to disseminate and legitimise their own agendas. As Catherine Needham (2004: 65) notes, governments can use processes of direct communication and consultation with their electorate to sideline the powers of elected legislatures to hold those governments to account. In their study of internet use in Russia, for example, Fossato and Lloyd (2008: 56) argue that while liberals may see new media technologies as tools for individual liberation their potential, in Russia at least, may primarily become focused upon social manipulation. Such scenarios have led the growing ranks of cyberpessimists towards what has been dubbed the reinforcement hypothesis. This theory is neatly summarized by Raab and Bellamy (2004: 21) when they argue that 'technology becomes a tool for the reinforcement of existing power structures.'

Martin Hand (2008: 77) has noted that western governments have grown increasingly interested in the use of digital technologies as tools for those governments' own reinvention and legitimization. One might suggest that any such reinvention appears to have been focused upon the entrenchment of power hierarchies rather than the opening up of government to greater transparency, accountability and popular participation; and one might also observe that the emphasis of most governmental initiatives in the deployment of these technologies has been concentrated upon the legitimization of extant institutions, structures and practices of power.

The uneven distribution of new media technologies' tools of access to intellectual and cultural capital is a problem not only for an electronic society as a whole; it is also a specifically critical issue for the practices of online government. Indeed, in 2003 the United Nations's World Public Sector Report, *E-Government at the Crossroads*, warned that broad sections of national populations were not necessarily reaping the benefits of state investments in electronic governance.

Neumann (2004: 208) pertinently asks whether we are creating a bipolar society split between those who have access to information technologies and an increasingly disempowered and disenfranchised sector of those who do not. According to Margolis (2007: 2) the electronic gap has come to reflect the socio-economic, educational and demographic divides of the non-virtual world, as the political and economic structures of cyberspace echo those found in the material world. As Bolter and Grusin (2000: 182) argue, access to these technologies confers a 'conspicuous social status' of some significance in the post-industrial world.

Nixon and Koutrakou (2007: xxi) report that individuals with lower levels of educational achievement are less likely to use the internet. This not only leads to a reinforcement of social, economic and educational schisms forged along demographic lines, but may also imply a set of specific and acute risks incurred by the adoption of new communication technologies for the purposes of political, parliamentary and governmental information, consultation and representation. As Nicholas Pleace (2007: 69) points out, within any nation sectors of society with low levels of internet access tend to correlate with those sectors which suffer low levels of household income (those very sectors which most need the engagement and support of the state): 'the implications for electronic voting are obvious. The poorer parts of the population are less likely to vote.' It may therefore appear inevitable that, as Åström (2004: 107) suggests, online voting 'shifts the bias toward the middle and upper classes: the already politically active.'

In an interview with Chris Middleton (1999), the IT ethicist Simon Rogerson proposed that 'the concept of online government implies literacy, and an awareness and acceptance of technology' but that the ability to meet these criteria for access was not universal. In this context, it is perhaps helpful to examine the case of a country which has stood at the forefront of developments in electronic democracy, government, industry and commerce: the small, modernizing and relatively new north-eastern European nation of Estonia, a country which ranked first in the world in Freedom House's 2012 index of internet freedom – coming second only to Iceland the following year.

E-stonia

Perched on the north-eastern edge of Europe, and wedged in between Finland, Russia and Latvia, the Baltic republic of Estonia achieved independence from the Soviet Union in 1991, and joined the European Union on 1 May 2004.

Stephen Coleman (2005a: 6) has commended the Estonian government for its practical uses of the internet to foster democratic participation. Pratchett (2007: 10) similarly cites Estonian examples of major e-democracy initiatives. Much has been made of Estonia's leap into the electronic age, perhaps because its rapid and massive adoption of new media technologies may be seen as affording a bridge from its past as a minor republic of the Soviet Union, annexed by Stalin in 1940, towards its current situation as an independent democracy based on the principles of market economics, and as a member of NATO and of the European Union. In many ways Estonia has represented to various commentators an admirable model for the development of liberal democratic nationhood in the wake of political oppression and a command economy.

In 2004 a ranking of the global development of e-government published by researchers at Harvard University placed Estonia in fifth place (Boyd 2004). The European Commission's fifth annual survey of online government services in Europe, released in March 2005, showed that, while most of the European Union's then ten newest member states scored in the lower half of the table – with average levels of e-government equivalent to those enjoyed by the EU's older member states in 2003 – Estonia alone of the new members appeared in the upper half (European

Commission 2005). Indeed, according to Ernsdorff and Berbec (2007: 171), 'Estonia stands as the e-government leader in Central and Eastern Europe and as third in the world in e-government systems.'

One of the more controversial aspects of Estonia's adventures in virtual administration was established in 2001 as the flagship project of its system of e-government. *Täna Otsustan Mina* ('Today I Decide') – or 'TOM' – was a website on which Estonian citizens could present proposals for legislation. If a proposal received sufficient support, it was to be discussed by the government. Although the portal boasted some 7,000 registered users, there were within a few years of its launch only 10 or 20 active members. The TOM portal prompted a number of minor changes in Estonian legislation and governance. One of these was a proposal to put the clocks forward in the spring and back in the autumn. Another was an amendment to the law on the possession of dangerous weapons – an exemption which permitted students of Tartu University (the country's oldest and most prestigious seat of learning) to carry swords on ceremonial occasions. The TOM portal received 359 proposals in the first six months of its operation, but by 2005 received only 49 proposals for the whole year (Ernsdorrf and Berbec 2007: 176). In fact, the website was generally considered an object of national ridicule, embarrassment or indifference. Mart Parve, technology correspondent for Estonia's most popular daily newspaper, *Postimees*, has called the initiative farcical, and has characterized its regular users as freaks and geeks: 'Some famous freaks are trying to start new laws, but it's not working. We've been quite pessimistic about the state since Soviet times. E-government is not employed at the level it should be. We're very interested in new technologies, but we don't use them properly.'

Estonia's TOM website eventually found itself sidelined and a new portal became tacitly acknowledged by the country's political elite as the primary arena for the proposal and discussion of legislation (Ernsdorff and Berbec 2007: 177). This website was created by an independent body, the Estonian Law Centre Foundation. Public opinion and legislative influence were now mediated neither by parliament nor even by a public or publicly accountable organisation.

On 4 June 2008 the Estonian government launched the *Osalusveeb* (Participation Web) as its successor to the TOM portal. According to the government website, this site would offer a chance for 'everyone [to] make suggestions to the state for simplifying public services.' With such an agenda set in advance, it seemed unlikely that this site would offer significant opportunities for the Estonian people to formulate policy alternatives to the principles of public sector downsizing so essential to the market economics of its centre right administrations. This new forum for debate appeared to be telling the Estonian electorate that their civic duty (and the extent of their democratic privilege) was to devise policy details appropriate to the government's own political programme. More recently the site's remit has welcomed 'efficient and relevant opinions and proposals.' It is unclear what the Estonian government feels such relevance and efficiency might comprise.

The use of citizen portals for online policy consultation is not of course an idea unique to Estonia. On 1 July 2010 the UK's Liberal-Conservative coalition government, for example, announced the launch of the 'Your Freedom' website, a platform from which – in the words of a *BBC News* report – the public would be able to 'nominate laws and regulations they would like to see abolished [and] also be able to propose ways to reduce bureaucracy.' Although this was reported by the BBC as being advanced as the biggest online policy-sourcing initiative by 'any government ever', one might note its similarities to the similar sites launched by the Estonian government over the previous decade – especially in terms of their propensity to promote policies designed to reduce the influence and cost of the public sector.

The coalition's Deputy Prime Minister Nick Clegg appeared on the BBC's *Breakfast* news programme on 1 July 2010 to propose, rather surprisingly, that this was, in some way, a comic exercise in the recognition of legislative absurdity: 'This could release a sense of fun as people think of silly rules that need to be scrapped.' Clegg added: 'I've just discovered for instance, would you believe it, that there's still an old law in the statute book that says it's an offence if you don't report a grey squirrel in your own back garden.' The absurdity of Clegg's example diminished the seriousness of the project from the outset. One might imagine that this initiative represented nothing more than a populist veneer of democratic consultation which offered the electorate an illusion of agency rather than a significant process of empowerment. After a month of this crowdsourcing policy initiative *The Daily Telegraph* reported, on 3 August 2010, that 'despite tens of thousands of public responses sent to various government departments, not one has shown a willingness to amend policy.' The *Telegraph* noted that, although the government had received responses in areas as diverse as health provision, pensions and taxation, transport and wildlife policy, each Whitehall department's official response appeared to interpret the public comments as 'an endorsement of existing policy.'

In August 2011 the same British government established a website to accept e-petitions. An attempt under the previous Labour administration had – according to Scott Wright (2011) – been hijacked by a relatively small number of serial petitioners. Wright (2012: 466) has observed that 'while there were a small number of cases where e-petitions influenced policy, the vast majority disappeared into a vacuum.' The petitions on this site had, for example, included one signed by 50,000 people in 2008 suggesting that motoring television present Jeremy Clarkson should be appointed Prime Minister. But in the second decade of the twenty-first century the new coalition government was of course promising something radically new.

On the day of its 2011 launch the coalition's latest website, which promised that any petition receiving more than 100,000 signatures would be debated in Parliament, had already fallen foul of eccentric minority interests. As *The Daily Telegraph* reported on 4 August 2011, 'right wing internet bloggers have been collecting signatures for several days calling for the reintroduction of the death penalty.' Later that day, the BBC noted that the return of the death penalty headed what it described as 'the list of demands.' The list also included the nation's departure from the European Union and limiting prison food to bread and water. The following week, however, the BBC reported that, in response to a series of riots which had swept the nation over the previous few days, the most signed e-petition on the website was one which called for convicted rioters to lose benefit payments. The petition, which had swiftly gained the necessary 100,000 signatories, specifically proposed that 'no taxpayer should have to contribute to those who have destroyed property, stolen from their community and shown a disregard for the country that provides for them.' One would assume that this rather blanket statement might suggest, somewhat beyond its original context, that not only social security benefits but also healthcare, education and other public services (including prisons, courts and the police) should be denied to anyone convicted of committing crimes against property. As such, it might appear as something of a manifesto for the creation and perpetuation of a permanent underclass. This mode of poorly considered, ill-informed and therefore fundamentally undemocratic demagoguery seems to represent the politics of the mob and of knee-jerk hysteria rather than the reasoned debate of a public sphere. As a *Guardian* newspaper editorial of November 2011 cautioned, 'this risks becoming mob rule.'

In fact the British government went on to do its best to ignore this e-petition and its demands were sidelined in the parliamentary response to the riots and were not put forward to be enacted into legislation. This appeared to demonstrate, yet again, that while new media may be useful in riling demagogic fervour to make the masses think their voices count, power elites continue to ignore those voices. Indeed, even such well-meaning online movements as the Invisible Children campaign for the arrest of the Ugandan guerrilla leader Joseph Kony have, as Harding (2012) points out, been criticized for their western-centred naïvety and their lack of useful impact – insofar as the mouse-clicking abilities of American college students do not represent a prerequisite for conflict resolution in Africa.

In August 2011, shortly after the success of the petition for rioters to lose their benefits, another e-petition appeared on the British government website – although it was removed the following month when someone in authority eventually got round to noticing it:

> All rioters and looters from the recent troubles in English cities should be banished to the Outer Hebrides for five years. This would be much, much cheaper than keeping them in expensive prisons, saving the taxpayer money. Five years of being forced to live in the Outer Hebrides with none of the comforts of English city living e.g. running water, electricity, decent food, culture and shopping, will put them on the straight and narrow, and frighten them not to riot or loot again. Many local people there look after sheep part-time, so they can earn a small amount of extra money looking after rioters and looters as well.

It is sometimes difficult to tell which are the sincere petitions and which are the hoaxes. In spring 2014 some of the trending e-petitions on the

UK government website included 'Tell Eric Pickles that allotments must not be sold off!' (having amassed more than 15,000 signatures), 'We oppose the proposal [sic] to change the August Bank Holday [sic] to Margaret Thatcher Day' (with more than 50,000 signatories and garnering the official response that 'there is no precedent for naming public holidays after an individual and the Government has no plans to do so') and 'Promote cycling by implementing the recommendations in the Get Britain Cycling report' (with more than 70,000 supporters and prompting the response that the 'Government takes cycling very seriously and is committed to leading the country in getting more people cycling, more safely, more often'). More than 5,000 signed a petition demanding the banning of the illegal release of otters – which are apparently not only posing the threat of 'ecological disaster' but which are also 'invading urban areas attacking ornamental fish in private garden ponds.' More than 18,000 signed a petition lamenting the fact that Coventry City football club had been relocated to Northampton. More than 45,000 people put their names to a petition calling on the government not to ban the Islamic face veil (and more than 5,000 signed another petition calling for exactly the same thing not to happen); the government responded that it had no intention of doing so.

It was reported in August 2012 that within a year of the launch of the website, 6.4 million signatures had appeared on 36,000 petitions – ten of which had reached the 100,000 mark necessary to trigger a parliamentary debate, and eight of which had by then been debated in parliament. However, a report published by the Hansard Society in May 2012 had warned that there was 'no agreement about the purpose of e-petitions', suggesting that it was uncertain whether there were to be viewed as merely 'an easy way to influence government policy' or whether they might 'be used to empower the public through greater engagement in the political and parliamentary process, providing for deliberation on the issues of concern.' The Hansard Society observed that the system was 'falling short of public and media expectations' and recommended that 'if the House of Commons is to be responsible for responding to petitioners' concerns then it should take over the running of the system from the government.' Two years later, in March 2014, the Hansard Society reported that the Leader of the House of Commons had 'indicated he is considering changes to the e-petitions system including the possible introduction of a Petitions Committee as recommended in our report.' It suggested that any such changes might not be introduced until after the next general election, in 2015.

Electronic democracy

On the other side of Europe, experiments in online democracy have been rather more radical and rather swifter to progress. In May 2005 the Estonian parliament passed legislation to introduce online voting at the country's local elections that October. According to *The Baltic Times* newspaper, 'Estonia would become the only country in the world where people could vote through the internet from home. Although online voting is widespread in other countries, a voter must conduct his E-vote at a polling station computer.' The Associated Press added: 'Voters will need an electronic ID card, an ID-card reader and internet access [...] It is estimated that nearly 1 million of Estonia's 1.4 million residents already have an official electronic ID card. The ID cards, launched in 2002, include small microchips and offer secure e-signing through a reader attached to their computers.' It should be noted that it was estimated that, at the time, nearly a quarter of Estonia's population did not hold these ID cards, and it was believed that these people were mainly minority Russian-speaking residents.

According to Aleksei Gunter, a leading journalist at the Estonian newspaper *Postimees*, 'at the 2005 local elections, most of the e-votes went to the Reform Party. That was somewhat predictable, because the young and the well-off, who obviously have the means and the interest to use new technologies, favour that party.' In fact, according to the government's own report on that exercise in e-democracy, the Reform Party gave out ID card readers to their supporters during the election campaign (Madise et al. 2006: 41). The report of the Estonian National Electoral Committee (2007) shows that nearly 35 per cent of the electronic vote went to the Reform Party – significantly higher than the 28 per cent which it achieved in the total vote. As Ernsdorff and Berbec (2007: 178) report, 'some political parties considered e-voting an opportunity to increase their support, while others conceived it a threat.' Indeed, the then President of Estonia, Arnold Rüütel, attempted to veto the legislation which permitted online voting, on the grounds of its inherently inequitable nature and its openness to fraudulent manipulation, but was eventually overruled by the country's Supreme Court.

On 28 February 2007 Estonia extended e-voting to its national parliamentary elections. In March 2007 the BBC announced that 'Estonia has become the first country to use internet voting in parliamentary elections.' *The Baltic Times* newspaper added that 'in what was hailed as the world's first full-scale internet election, a total of 30,243 voters chose to log their votes online' (Alas 2007a). The total number of votes cast at the election was 550,213 and the total number of eligible voters circa 940,000. Ernsdorff and Berbec (2007: 171) have written that Estonia 'is setting an example in e-democracy throughout the European Union, being the first country in the world to enable all its citizens to vote over the internet in political elections.' One might, however, call into question Ernsdorff and Berbec's use of the word *all*: it has been estimated that only about three-quarters of Estonians use the internet.

Voter mobilisation is, of course, a key factor in the winning of elections (see Oberholzer-Gee and Waldfogel 2005: 74; Trost and Grossmann 2005: 128). In facilitating the voting process for their typical supporters, political parties more likely attract the most e-literate portion of the electorate (in Estonia's case, those of the centre right) would therefore be afforded an obvious electoral advantage by this mode of electronic democracy. Ülle Madise (2007), Director of Audit at Estonia's State Audit Office, has claimed that online voting offers no advantage for e-voters – but, if that indeed were the case, one wonders why the state would bother with the trouble and expense of it at all. As Trechsel (2007: 37) points out, nearly 86 per cent of Estonians who chose to vote online did so because they found it more convenient than by traditional methods.

In an otherwise remarkably optimistic essay, Ernsdorff and Berbec (2007: 178) admit that 'e-voting has never been the result of popular demand but rather a result of the imposition of yet another initiative by a young Estonian political elite.' This is a position with which Jaak Aab,

Estonia's Social Affairs Minister at the time of Estonia's first experiments in electronic democracy, would strongly disagree. Aab, who held this ministerial portfolio from April 2005 to April 2007, has commented:

> I believe that online voting encourages people to take an active part in democracy, because it gives an additional possibility to vote. It is especially important for people who cannot or have lower motivation to go to vote to the designated voting place. I am not concerned that it may primarily encourage participation among the educated middle classes, because Estonia doesn't have a big gap in internet use, as lots of other European countries do. The Estonian government's aim is to provide the internet to all Estonian people.

However, there appear to be several problems with Aab's argument. His first assertion – that online voting promotes democratic participation – seems to contradict the evidence of various empirical studies. Gibson et al. (2004: 3), for example, cite research which suggests that no modes of internet use have been found to have any significant effect on individuals' tendencies to engage in politics.

Aab's dismissal of the education gap and the technological divide as irrelevant to Estonia is extraordinary, not only in the light of the country's massive socio-economic divisions, but also because it specifically contradicts the findings of his government's own report on the 2005 elections: that there were more people with high levels of education among e-voters (Madise et al. 2006: 30). Breuer and Trechsel's *Report for the Council of Europe* (2006: section 7) on the 2005 elections meanwhile found that Estonia's e-voting opportunities proved relatively unattractive for the elderly, for those with limited IT skills and for minority language speakers. We may note in this context that in 2010 the United Nations Committee on the Elimination of Racial Discrimination noted the low level of political participation of ethnic minorities in Estonia. (Indeed the OSCE report on Estonia's parliamentary elections of March 2011 observed that, in contravention of OSCE requirements, Estonia's 'long-term residents with undetermined citizenship do not have the right to join political parties.')

The *Report for the Council of Europe* on the 2007 election noted that 'the share of highly educated voters was almost 20 percentage points higher among e-voters than among traditional voters' (Trechsel 2007: 43).

Trechsel's report stresses that 'the highest-income category is heavily over-represented among e-voters' and that 'a very large part of the Russian speaking community [refrained] from using this tool' (Trechsel 2007: 44, 6). Trechsel's study demonstrates that more than 62 per cent of voters who elected not to vote online did so because they lacked the necessary facilities (Trechsel 2007: 38). It also points out that 'e-voters do not only differ [from traditional voters] with regard to their socio-demographic and economic profiles, but they also do so [...] with regard to their political preferences' (Trechsel 2007: 49).

To the extent that socio-economic status can be elaborated upon geographical lines, it seems significant that, with the exception of the university city of Tartu, the areas of Estonia which, according to the report of the Estonian National Electoral Committee (2007), demonstrated the highest use of electronic voting among the national turnout were Tallinn itself and its neighbouring counties in the affluent north-west of the country. With barely more than a third of Tallinn's score, the economically depressed county of Ida-Viru on Estonia's eastern border with Russia showed the lowest rate of adoption of the electronic system.

The Organisation for Security and Co-operation in Europe's report on Estonian e-voting went so far as to question 'whether [in future] the internet should be available as a voting method, or alternatively whether it should be used only on a limited basis or not at all' (OSCE 2007: 2). This lack of enthusiasm has not, however, prevented Estonia from forging ahead with this system: in the European elections of June 2009 58,669 Estonians voted online, while Estonia's municipal elections of October 2009 saw 104,413 people (9.5 per cent of the eligible electorate and 15.7 per cent of those who voted) vote online. By the national election of March 2011 140,846 people (24.3 per cent of those who voted) voted online.

In October 2007 it was reported that Estonia planned to introduce an e-voting mechanism involving the use of mobile telephones (Alas 2007b). Despite initial hopes that direct mobile telephone voting would be implemented in time for the municipal elections of 2009, in December 2008 the Estonian parliament ratified plans to implement this system for the first time at the parliamentary elections of 2011.

In Estonia's 2011 elections, nearly a quarter of those who voted did so online – including those who were (for the first time) permitted to vote by mobile phone. Kitsing (2013) has however argued that there is still no evidence that Estonian e-voting increases turnout, but suggests that it has resulted in 'reduced civic engagement' as well as 'privacy and security concerns.' Kitsing has added that (yet again) 'older and less educated segments of the population can experience significant barriers in exploiting internet voting' – and, as a result, the system tends to favour the centre right parties (who continue to champion its roll-out) rather than the more populist end of the party political spectrum.

The shape of things to come?

One of the problems repeatedly encountered by the idea that developments in media technologies and formats can be effectively harnessed for the promotion of democratic participation is that these tools are for the most part employed by governments, political parties and political and commercial interest groups not for the sake of the reinvigoration of the public sphere but with their focus fixed resolutely upon the reinforcement of their own specific bases of power. Thus, as Blumler and Gurevitch (1995: 221) have warned, attempts to manage public communication in order to manipulate the public have resulted in the alienation of citizens from democratic processes – the very trend of political exclusion which such strategies might ostensibly be intended to reverse.

The Estonian example demonstrates, if nothing else, that technological developments do not necessarily result in greater levels of participatory citizenship, democratic accountability or social justice. Indeed, any attempts to deploy these technologies to advance democratic values have been undermined by the failure of the nation's e-visionaries to see that, rather than solving the country's social problems, the imposition of new technologies upon relatively youthful processes of government and democracy may in fact exacerbate those problems.

In a study online political participation in another post-Communist nation, Koc-Michalska, Lilleker and Surowiec (2013: 98–99) argue that although there is evidence that 'a small elite appear to be building Poland's online political sphere' the fact that these agenda-setters represent a number of distinct and competing elites suggests that this diversity of usage may eventually come to propagate the extension of 'structures of political accessibility.' Visions of the prospects for such a participatory expansion remain at best, however, ambivalent. It is clear that we cannot look to social media alone to provide the answers.

Even the founding father of the Estonia's digital revolution, Linnar Viik, has expressed doubt over his country's experiments in electronic democracy. In 1997 Viik returned to Estonia from his studies in Finland bursting with ideas to promote his country's electronic development, and became the government adviser who changed the face of Tallinn's economy. More recently, however, Viik has argued that 'e-democracy doesn't have a real impact on the democratic process. Democracy in Estonia is like a small child. I can compare it to my five-year-old son. He can talk, he knows some manners, he knows how to pee – but he's still learning. This technology is just a tool.'

Yet Viik believes that traditional modes of parliamentary democracy are now obsolete: 'I don't so much believe in representative democracy. I believe in participatory democracy.' Viik's distrust of traditional political structures and processes and his hope that new media technologies will foster democratic participation reflect the view of such studies as those of Coleman and Spiller (2003) and Coleman and Blumler (2009), a view that new media technologies have a crucial role to play in any strategy which hopes to restore public trust and participation in democratic processes. However, for as long as such participation remains limited by educational, demographic and socio-economic conditions, this mode of democracy will continue to be dogged by the concern that it has become the instrument of entrenched power structures and privileged interest groups, and that it offers only the illusion of participation in place of real democratic empowerment.

This illusion of empowerment recalls that which Henrik Bang (2010: 261) describes as a process which seeks at once to empower and to *domesticate*. This mode of governance, writes Bang (2010: 247), 'aims at empowering as many people as possible; not primarily for their own sake but to get them to help their organisations to get the minimal wholeness, coherence and effectiveness that they need.' It attempts to 'gain its autonomy as a mode of rational rule via its own specialised semantic, which [...] has the consequence of ignoring those everyday narratives and forms of life that do not possess this kind [...] of specialisation' (Bang 2010: 250). Such empowerment is thus empowering only for those willing and able to fit its model.

On 10 April 2014 – on the first day of that year's national elections in the world's largest democracy – the BBC's *Today* programme observed that India's more than 800 million eligible voters had access to an e-voting system, albeit notably one which still required the physical presence of the voter in the polling station: 'in India electronic voting has reached even the most far-flung areas – so it's not a case of ballots in the boxes so much as buttons on the voting machines.' Yet the Indian system – launched in the early 1980s – has been dogged by allegations of misuse and malfunction: indeed the *Economic Times* had reported on 5 April 2014 that machine faults had prompted the random checking of voting machines. At this end of the spectrum, it seems then that the jury remains out on the question of the advantages to the development of democracy offered by the deployment of such digital information and communication technologies. On a smaller scale too, the impacts of these technologies remain uncertain.

Gibson et al. (2014) have suggested that 'parochial' forms of civic participation – 'individualistic activities which aim to solve a particular or private issue' – might offer 'the potential to act as precursors to other forms of civic and political engagement' but it remains unclear whether this is in fact the case. They have suggested that the citizen who successfully convinces her local council to solve a particular local issue which directly impacts upon her own life may as a result develop stronger ties of democratic trust and therefore of political participation. Their initial investigation into this possibility has however suggested that those armchair activists who pursue the agendas of such civic self-help websites as fixmystreet.com, writetothem. com and theyworkforyou.com (or at least those who responded to their surveys thereon) are 'already very engaged' in such civic activities – indeed only 17 per cent of their study's respondents were first-time users of those sites. It is also uncertain whether such parochial and consumerist activities would promote or, conversely, sublimate impulses towards broader political commitment and participation offline. The fact that more than 30 per cent of their respondents were retired suggests, as they have said, that this sample was not representative of the population nor indeed of internet users in general; but also may suggest that this constituency may eventually prove rather less likely to engage in long-term political action. In these terms, parochial participation may be seen perhaps as a substitute for broader political engagement which further defers such engagement because it gives a mere sense of engagement and therefore dissolves the frustration born of disengagement (a frustration which may nevertheless prompt engagement).

Jackson et al. (2014) have argued through their study of 'how political discussion emerges from non-political spaces' online that political talk on such apparently apolitical British websites as Moneysavingexpert, DigitalSpy and Netmums may 'lead to political action.' Their survey of more than a million posts on these sites has shown how 'political talk emerges throughout everyday conversation' and that this 'overlap between the personal and the political' may promote the mobilization of political activity. They have however stressed that new media platforms do not in themselves hold the secret to a renaissance in political engagement; but that – although they discovered that these discussions tended to remain 'remarkably UK-centric' (featuring 'barely anything on world events') – they might offer some potential for the longer-term sponsorship of more broadly based modes of political engagement.

Figures published by the UK's Office of National Statistics showed that in 2013, while 66 per cent of British people used the internet to find information on goods and services, and 55 per cent used it for social networking purposes, only ten per cent used it to post their views on civic or political issues, and only seven per cent used the technology to take part in online consultations or formally register their support for civic or political issues. It may be supposed that all revolutions must first be generated locally (individually and parochially) and at the level of the grass roots; but it may also be argued that the notion of a consumer revolution is a contradiction

in terms: that, as Scullion (2013: 64) has suggested, 'the meanings attributed to consumer choice-making serve to nullify much of the salience of
political engagement, and, in the process, create the somewhat peculiar
mixture of an often personally belligerent, yet collectively conservative,
contemporary electorate.' It seems unlikely that new media technologies
will in themselves be able to resolve these contradictions, and that if we
think that they do (if we believe that they have radically transformed the
terms of socio-political engagement) they will be even less likely to do so.

That then continues to be the problem: not that these technologies
do not empower us (of course they don't; of course they couldn't possibly
do so) but that we think they do. This, as we shall see, appears to apply
in the fantasies of virtual play as in the dreams of electronic democracy.

CHAPTER 3

War Games

We have been experiencing, for half a century, a conflation of material history and its electronic mediation, and this phenomenon is perhaps at its most remarkable in the conduct and representation of military conflict.

Jean Baudrillard (1988: 49) wrote of Vietnam as a television war – but Vietnam also of course eventually became a cinematic war, a war primarily recalled in the popular imagination by such films as *The Deer Hunter*, *Apocalypse Now*, *Platoon* and *Full Metal Jacket*. Another postmodern conflict, Operation Restore Hope, America's vain attempt to bring order to Somalia in 1992–1993, also began as an event staged for the TV cameras (even to the extent that the Pentagon is said to have consulted CNN on the scheduling of the U. S. landings in Mogadishu), and ended up as a film by Ridley Scott: a five-month military debacle immortalized as *Black Hawk Down* (2001).

The BBC's World Affairs Editor John Simpson's declaration of his personal liberation of Kabul in November 2001 and Donald Rumsfeld's announcement in February 2006 that newsrooms had become crucial battlefields in the War on Terror are two well-known examples of the convergence of media and military perspectives. As Baudrillard (2005: 77) wrote, 'media and images are part of the Integral Reality of war.' Ronald Reagan's abortive *Star Wars* programme stands as a landmark moment in this process, and that Hollywood President's Tinseltown apocalypse was to be echoed in George W. Bush's cowboy diplomacy and in his administration's use of popular filmmakers as strategic imagineers.

The Oscar-winning director Kathryn Bigelow was already working on a film about the hunt for Osama Bin Laden – the film that then became *Zero Dark Thirty* (2012) – when Bin Laden's killing was announced. In August 2011 Bigelow denied Republican claims that the filmmakers had been fed information by the Obama administration. An American government report published in June 2013 suggested that classified information relating to the Bin Laden operation had been discussed by a former director of the CIA at an event at which the film's screenwriter had been present.

The relationship between Hollywood culture and real-world military activities was underlined by claims made in 2011 that Guantanamo Bay interrogators had copied torture techniques from the TV series *24*. A similar confusion of military reality and fantasy entertainment took place that same year when Britain's ITV broadcast footage purportedly of IRA military training that turned out to be from a video game. As Jeffries (2011) reported, Marek Spanel, chief executive of the game's developer Bohemia Interactive Studio, said that he considered this 'a bizarre appreciation of the level of realism' incorporated into the game. Such confusion is hardly unique to this incident. The following May, for example, a BBC news report on the United Nations Security Council mistakenly showed the logo for the United Nations Space Command from the computer game *Halo*.

At the height of the media hype surrounding the early days of the War on Terror, the overlap between the media and military power had appeared to become increasingly overt. The jingoistic and war-mongering stances of much of the British and American tabloid recalled Orson Welles's immortal exhortation to a journalist in *Citizen Kane* (1941) – paraphrasing what has been (possibly apocryphally) reported as an 1897 declaration from real-life newspaper tycoon William Randolph Hearst: 'You provide the prose-poems. I'll provide the war.'

D. W. Griffith's 1914 film *The Life of General Villa* had established a model for media-military synergy by directly funding the Mexican Revolution – which it documented through a mixture of authentic footage and fictionalized scenes. We may not quite have reached the situation suggested by the 1997 James Bond film *Tomorrow Never Dies* in which Jonathan Pryce's media mogul attempts to start a war between Britain and China in order to boost newspaper sales and penetrate new satellite broadcasting markets; yet we are growing ever more acclimatised to that absurdity. Indeed in 2012 Tony Blair's former Director of Communications and Strategy Alastair Campbell published claims that in 2003 Rupert

Murdoch had participated in an attempt to push Blair into accelerating British involvement in the planned invasion of Iraq.

Slavoj Žižek (2002: 15) has proposed that the events of 9/11 itself may be seen as representing the physical manifestation of globalized media product – a Hollywood catastrophe movie: a further blurring of material reality and popular entertainment – in the words of Tumber and Webster (2006: 4) a *media event*. Within five years, the events of 11 September 2001 had already lent themselves to a number of film adaptations including Oliver Stone's *World Trade Center* (2006) and Paul Greengrass's *United 93* (2006) – and had even prompted a comic book version, Sid Jacobson and Ernie Colon's *The 9/11 Report: A Graphic Adaptation* (2006), a video game called *9–11 Survivor* (2003), and, in 2011, a colouring book for children entitled *We Shall Never Forget 9/11: The Kids' Book of Freedom*. In an article published in *The Times* on 12 September 2001, Michael Gove had written that that 'the scenario of a Tom Clancy thriller or Spielberg blockbuster was now unfolding live on the world's television screens.' The events of that day were especially reminiscent of one particular Tom Clancy blockbuster, a novel entitled *Debt of Honour*, a book which climaxes with a terrorist crashing a civilian airliner into Washington. (It seems no coincidence that CNN chose to interview Tom Clancy during its live coverage of the attacks on the WTC.) More extraordinarily, six months prior to the catastrophe of 2001, the debut episode of the science fiction TV series *The Lone Gunmen* had depicted a terrorist plot to fly a hijacked airliner into the World Trade Center itself.

The self-styled enemies of the West are inscribed within its pandemic media culture. Saddam Hussein reputedly enjoyed Scottish football as he ate Jaffa Cakes in his prison cell, but did not live to see his 2000 novel *Zabibah and the King* adapted into a 2012 satire by filmmaker Sacha Baron Cohen; Kim Jong-il was reported to be an aficionado of Hollywood cinema (he once boasted that he owned 'all the Academy Award movies'); indeed, according to his former lover Kola Boof's *Diary of a Lost Girl* (2006), Osama bin Laden himself was a big fan of *Miami Vice*, *The Wonder Years*, Whitney Houston and *Playboy*. As Slavoj Žižek (2008: 73) has suggested, 'the fundamentalists are already like us [...] secretly, they have internalised our standards and measure themselves by them.' Hence these pseudo-fundamentalists' apparent obsessions with certain western media texts: from Salman Rushdie's *Satanic Verses* and the *Jyllands-Posten* Muhammad cartoons, through the reality TV show *The X Factor* and the cartoon TV series *South Park*, to the Islamophobic films of Theo Van Gogh, Geert Wilders and Nakoula Basseley Nakoula (all of which have prompted death threats or actual murders).

Between 2007 and 2009 a Hamas-sponsored Palestinian television station ran a children's series entitled *Tomorrow's Pioneers* which featured a character called Farfur, a zealot who taught the children of Palestine of their mission to 'return the Islamic community to its former greatness, and liberate Jerusalem, God willing, liberate Iraq, God willing, and liberate all the countries of the Muslims invaded by the murderers.' Farfur was a giant rodent, one who bore a very striking resemblance to Mickey Mouse. Even terrorism, it seems, has become disneyfied.

Years before 9/11, Jean Baudrillard (1994: 21) had written that 'all the holdups, airline hijackings, etc. [...] are already inscribed in the decoding and orchestration rituals of the media.' Or, as Slavoj Žižek (2002: 146) has put it: 'Jihad is already McJihad.' All the world's a game.

Gameworld

Material history ebbs away, to be replaced by the virtualizing culture of the mass media. This sense of the slippage of the material beneath the mediated is famously elaborated in Jean Baudrillard's *The Gulf War Did Not Take Place*, in which the French sociologist suggests that the Gulf War may be read not as a piece of real, material history but as a media event, a pseudo-event performed for the media. It is not perhaps insignificant that the Gulf War's emotionally sterile images of bombings – images captured by cameras mounted on warplanes, pictures which saturated the television coverage of that conflict – not only exposed a blurring of material-historical and electronically mediated perspectives, but also conspicuously translated acts of mass destruction into the visual idiom of the video game.

Andrew Darley (2000: 31) has argued that the primary benchmark of success in digital game design is the game's graphical verisimilitude, its representation's approximation to external reality. Yet one is tempted to suggest that Darley's argument might be inverted: that the verisimilitude of material reality may now conversely be judged by its approximation to the virtual world. Media texts do not merely reflect reality; as John Fiske (1987: 21) suggests, they construct it. The photographic realism at which the digital game has aimed is – as Bolter and Grusin (2000: 55) point out – precisely that: a realism whose model is not material reality itself so much as the visual perspectives of cinema. And, if the visual realism of the video game has been defined by another medium (if it has seen – because we have all come to see – the visual identity of cinema, with its sutured edits, close-ups and lens flares, as the benchmark of the real), then why should we not see the graphic idiom of the computer game as a new standard for the representation and perception of material reality? Egenfeldt-Nielsen et al. (2013: 7) have proposed that 'the video game's explosive evolution of creative possibility is beginning to influence significantly other types of expression' – for example, that 'movies and games are borrowing from each other's arsenals' – and have gone on to suggest that there are thus now generations for whom 'games are crucial' to 'the way in which they conceive of the world.'

Could we therefore not see that in what Jameson (1991:48) once called the *society of the simulacrum* the digital game has become a crucial yardstick for the real? If, as Katherine Hayles (2000: 69) has suggested, the citizens of late postmodernity live increasingly virtualized lives, then could it not be expected that we might begin to witness a situation in which, in the words of Geoff King and Tanya Krzywinska (2006: 200), 'the distinction between reality and simulation might occasionally appear to blur, like something out of the pages of Jean Baudrillard'?

It is not just that the virtual and the non-virtual are becoming indistinguishable; what is significant is that the non-virtual is increasingly subordinated to the virtual. This has been going on for quite some time. As early, for example, as 1995, Heim (1995: 68) had talked of 'the actual, non-virtual world.' In a 2002 essay on *Counter-Strike*, Wright, Boria and Breidenbach also tellingly referred to 'the non-virtual world.' In October 2005 Jeffrey

Zaslow published a feature in *The Wall Street Journal* in which he explored strategies by which parents might encourage their offspring to 'connect to the non-virtual world.' More recently, Holmes (2013: 160) has noted (as our last chapter suggested) that notions of civic status differ between online and 'non-virtual world forms of politics.' López et al. (2013: 11) have meanwhile observed that (as this chapter will also contend) 'intensive users of the virtual space find that it has changed the nature of their existence in the non-virtual world.' Further to that, Warf (2013: 39), and indeed this book as a whole, have argued that 'cyberspace reflects all of the inequalities and social divisions that permeate the non-virtual world.'

This phrase – 'the non-virtual world' – signals the prioritization of the virtual: the virtual is no longer the 'non-real'; the virtual is not defined by its relation to the real – the real is defined by its relation to the virtual; the real is now merely the 'non-virtual', a category of secondary significance. Material reality has been designated as merely the 'offline' world: in other words, it is defined as that which is not online (the primary reality). In April 2014 Facebook's Sheryl Sandberg told *The Guardian* that even 'the word "online" is becoming an old-fashioned word because we're all going to be connected all the time.' The digital game – like the internet itself – comes therefore for some to represent their first version of reality. Apparent improvements in games graphics (the narrowing of the gap between representation and reality) are therefore not only a result of the virtual having become more real: they may also result from the real having become more virtual.

Jane McGonigal (2008) has announced that lessons learned from the development of digital games could usefully be applied to the material world – 'not [to] make our games more realistic and lifelike, but [to] make our real life more game like.' This is a point she makes extremely clear in her 2011 bestseller *Reality is Broken: Why Games Make Us Better and How They Can Change the World*.

McGonigal has predicted that by 2023 a games designer will be dominated for a Nobel Peace Prize. In March 2010 the World Bank launched a digital game designed by McGonigal: entitled *EVOKE*, it promised 'a ten-week crash course in changing the world' by offering to 'empower people all over the world [...] to come up with creative solutions to urgent social problems.' But the video game can, it appears, save the individual as well as the world: in a TED Global talk of June 2012 McGonigal suggested that playing video games would allow you to extend your lifespan to 'live a life truer to your dreams' – and that 'with ten extra years you might even have time to play a few more games.' The gameworld is thus gaining existential primacy: the gameworld, for many, already *is* the world.

Despite McGonigal's enthusiasm, it is not always clear whether the gamification of western civilization is necessarily a good thing. Juul (2013: 10) has, for example, argued that 'the 2008 financial crisis was caused in part by large banks and financial institutions making their organizations too gamelike by giving employees the clear goal of approving as many loans as possible and punishing naysayers with termination.' Yet, despite such possible consequences, it is undoubtedly the case that video games have continued to become increasingly influential upon the evolution of late postmodern culture. As Juul (2012) has shown, mobile technologies have propagated broader audiences for digital games and have thus begun to reinvent contemporary populations as the players of such games – at a time when, as Newman (2013: x) has put it, 'games appear to be more prevalent (dare we say pervasive?) than ever.'

The dogs of war

If games may come to take over our lives, one is reminded in this connection of that avatar of virtual existence, 'FPS Doug', an obsessive digital gamer in the cult video series *Pure Pwnage. Pure Pwnage* was a series of low-budget mockumentaries which fictionalized the lives of the digital games players of Toronto. The series ran online for 18 episodes over two seasons from 2004 until 2008, before moving to Canada's Showcase cable television channel for a further eight episodes in 2010. Production on a film version started in 2013. The show's title (pronounced 'pure ownage') refers in gaming slang to a player's mastery of their game.

In the show's fifth episode a character called FPS (First Person Shooter) Doug notes: 'Sometimes I think maybe I want to join the army. I mean it's basically like FPS, except better graphics.' But is it not increasingly the case that, insofar as we view the material world through prevalent visual paradigms, the reality of military experience resembles the video game, albeit with somewhat *less* convincing graphics?

Pure Pwnage's lead character, games addict Jeremy similarly sees the world through the perspective of the video game. In the premiere episode he asserts the irrelevance (indeed the existential inferiority) of the non-virtual world: 'normal people wake up in the morning and they watch CNN. I just don't like that because that's kind of fake.' For Jeremy, the world of the news has become less real than the realm of the role-playing game. When in the sixth episode he discovers *World of Warcraft*, he announces that he has found 'a new place to live' – that he is 'going to live in Azeroth.'

One of the recurrent and defining characteristics of the superhero genre is that, when not assuming the mantle of their superheroic alter ego, the protagonist leads a normal human life – a life which grounds them in everyday morality – while the genre's villains tend to be caught within their surreal personae (rarely does one see the Joker, for example, bereft of his make-up and attending to his day job). In this way, these fables may act as warnings against an addiction to an alternative selfhood – when, as Jim Carrey discovers in *The Mask* (1984), or as Robert Louis Stevenson's Henry Jekyll and H. G. Wells's Invisible Man had found a century earlier, the metahuman self corrupts and overwhelms the original self. This danger of irreversible transformation is clearly addressed at those moments when superheroes choose to reject their assumed personae in order to reassert their humanity – as when, for example, Michael Keaton rips off his mask towards the end of Tim Burton's *Batman Returns* (1992). Yet the humanly impotent self is for the most part perhaps inevitably subsumed to the super-humanly powerful self, even when (as for the digital gamer) that super-human power is transitory and illusory. If all subjectivity is performative, and if that performativity is determined by external parameters, then an imaginary super-agency is clearly more attractive than the reality of mortal struggle and near-powerlessness. Faced with an aggrandizing virtual exist-ence whose swift seamlessness offers little room or reason for critical self-reflection, the gamer may thus all too easily succumb to the temptations of fleshlessness – the real desert of the real (that is, the desertion of the real), a fate which Joe Pantoliano's character Cypher seeks to embrace in the Wachowski brothers' *Matrix Reloaded* (2003). As *Pure Pwnage*'s Jeremy announces in the series's eleventh episode, 'in the future we're going to be living in tombs – we're just going to have stuff hooked up to our brains – and we're basically just going to sit round playing games all day long.' The Wachowski brothers' dystopia is Jeremy's utopia. Jeremy can no longer distinguish between virtual reality and real life, and no longer wants to: 'real life' is merely 'RL' to him, an inferior alternative to VR: 'RL is a game.' As one games enthusiast says in the second episode of the third season: 'just because something's virtual doesn't mean it's not real.'

Pure Pwnage producer Davin Lengyel has pointed out that the series's game-obsessed characters were all based on real people – either exaggerated versions of the actors and writers themselves or the people they play games against. He has, however, gone on to express some concern as to how much these games may influence their players:

> I've personally been influenced in my personal behaviour by games I play. I noticed this in particular with *Grand Theft Auto 4*. I'd play six or seven hours a day, and when I walked outside and I'd see a car of the same model as in the game, I'd have a desire to steal that car. If you practise a behaviour all day, it becomes natural to you. If I play this game for months, I'm going to be influenced by this game. You do spend a lot of time with your games – it's what you think of when you're at work – you want to get home to play your games.

Lengyel is particularly concerned by the way in which the military have used games for the purposes of indoctrination and recruitment:

> The U. S. military has developed training simulations which are kick-ass video games. But to use video games as a means for recruiting for the armed forces for me is ludicrous. They're using a medium of entertainment to create a positive brand for the U. S. military and to desensitize people to the dangers of combat. There's a new version of a flight simulator, sponsored by the U. S. Air Force, all their models provided by the Air Force – their slogan is 'This time it's real.' It makes me really sad when I hear soldiers saying it was very much like a video game when they were in combat. The learned behaviour is you can get shot and still win. It scares the hell out of me that this might be a learned behaviour that people might take into a combat situation. The soldiers in combat have to remind themselves what they're doing and that it's actually real – and that the person they've just shot was a real person.

Lengyel also notes one particular area of confusion between real-life conflict and the digital game: 'People who pilot drones – they're in the States – they shoot and kill people and fly the drone with a joystick – it's not dissimilar to a video game.' Game/life boundaries are clearly being blurred. Citing an interview with one such drone pilot, Lengyel has observed that the pilot had commented that the emotional impact and immediacy of this mode of warfare derives primarily from its generic and morphological familiarity, that it looks and feels just like a video game: 'killing a human is so easy and so familiar – like a video game.' The game has become the yardstick of reality; the virtual map has not only superseded but appears now to pre-empt the physical territory. We do not feel the emotional and moral significance of another's life and death in that they are of the same biological species, the same flesh and blood, as us – but only insofar as they are incarnate avatars of beings from a digital realm.

In the last episode of the *Pure Pwnage* TV series Doug's gaming reality finally overtakes his material existence, as the devotee of the FPS game *America's Army*, recruited at a gaming convention, enrols in the real United States military: 'what kind of adventure is this – RPG style?' It is difficult to see this as an act of empowerment. 'I'm not an idiot,' he says. 'Sure, I'll die a few times but there's no way I'll be dying as much as them.'

In March 2003 the UK's *Sun* newspaper had used a gaming metaphor when it ran on its front page a photo of a British soldier brandishing his gun beneath the headline: 'Game over: Blair tells Saddam his time is up.' Three days later the invasion of Iraq began.

The distinctions between the video game of the war and the war of the video game have significantly blurred. In 2002 the United States military launched the First Person Shooter game *America's Army*. The game was first intended as a training tool, but was swiftly repackaged and made available for free online for propaganda and recruitment purposes. By December 2011 the game's website boasted that it had served more than 12 million registered users. A press release of August 2013 described *America's Army* as using 'innovative technology to provide authentic and entertaining Army experiences that reflect the lives, training, technology, skills, careers and values of a United States Army Soldier.' The distinctions and the tensions between authenticity and entertainment are typically blurred.

In 2008 the website for *America's Army* had promoted its second edition, *America's Army: True Soldiers*, as 'the only game based on the experiences of real U. S. Army soldiers.' This version of the game announced that it had been 'created by soldiers, developed by gamers, tested by heroes.' It is notable that this promotional copy recognized no final distinction between soldiers and gamers: both are *heroes* – neither of them *play* the game; they both *test* it. This testing serves multiple functions: the reality of the simulation is tested by soldiers and gamers alike to temper and strengthen its military value as a training tool, while that simulated reality thereby becomes the dominant version of perceived reality, and thus works as a tool both for propaganda and for (ideological and actual) military recruitment.

When *America's Army 3* was launched in June 2009, it continued to stress its proximity to the real: 'Your journey will begin with training so realistic you'll swear you're actually there. That's because the training in *AA3* was created by the U. S. Army and is based on real Army training. In *America's Army 3*, characters are more authentic than characters in any other video game. From the way the weapons look to how they are deployed and how they sound, the level of realism is unparalleled.' The website was also updated to include profiles of real soldiers serving in the U. S. military awarded for bravery; their integration into the paratext of the game again emphasized not only the verisimilitude but in effect the reality of the virtual experience: 'The opportunity to meet these Real Heroes is one of the many ways that *America's Army 3* is a game like no other.'

Launched in August 2013, its fourth iteration *America's Army: Proving Grounds* promised to develop in its players 'the full spectrum of skills and roles you'll need to support future operations.' It is not made explicit whether these future operations are expected to take place in the game-world or in the real world. That ambiguity might appear quite intentional. A link takes users from the *America's Army* website to goarmy.com: a site visually redolent of the *America's Army* webpages, and one through which the user can join the U. S. military and possibly even qualify for a '$20,000 enlistment bonus.'

Zhan Li (2004: 137) has suggested that *America's Army* represents an ambiguous space caught between the political, the military, the commercial, the virtual and the material. This blurring of traditional generic,

ontological and epistemological boundaries is perhaps most strikingly performed by *America's Army*'s touring recruitment circus, a disconcertingly
physical roadshow – the chance to see some 'real' U. S. military hardware,
alongside videos of serving soldiers and, of course, at the forefront of all
this (as the prioritized mode of reality), the gaming experience itself. This
Virtual Army Experience (VAE) and its 'mission simulator area' provide
participants 'with a virtual test drive of the Army.' There is something
disappointingly incongruous in the physical aspect of the VAE's recruitment roadshow; its physical presence seems to represent a poor copy of
the digital game itself. This phenomenon is hardly unique to *America's
Army*: when, for example, *World of Warcraft* fans don their home-made
costumes and enact gameplay performances at their conventions and other
such gatherings, are not these materializations of their gaming somehow
less realistic, and therefore (to them) less real, than the virtual world they
attempt to emulate?

Re-establishing the primacy of the virtual over the material, the
America's Army website has offered a virtual online tour of the VAE – a
simulation of a simulation, one which leads the eye through a computer-
graphic reconstruction of a room full of computers, their screens displaying
scenes from the original game. This virtual tour does not allow the user
any navigational control: like a gameplay video, it leads its impotent viewer
through its environment – towards the inevitable end of interpellation and
recruitment: to sign up to the U. S. military or at least to its ideological
perspective. The VAE's virtual tour represents a meta-simulacrum – an
electronic simulation (a virtual tour) of a physical simulation (the VAE
touring event) of an electronic simulation (*America's Army* – the game)
of a military reality which is increasingly virtualized.

This process of virtualization has also been seen in the conduct of war
itself. Warfare has begun to adopt the characteristics of the digital game.
King and Krzywinska (2006: 199) have pointed out that material warfare
is increasingly mediated by such devices as head-mounted displays offering visuals highly reminiscent of video games. In the realm of fantasy, the
2014 CBS television series *Intelligence* imagined a super-soldier whose
brain was connected directly into the internet; but in reality we are not, of
course, quite there yet. Not *quite*: on 10 October 2013 the BBC reported

that the U. S. Army was working to develop 'revolutionary smart armour that would give its troops superhuman strength.' The BBC added that this '*Iron Man*-style suit' would include 'wide-area networking and a wearable computer similar to Google Glass.' (Cf. Waterfield 2014.)

The super-soldier of the future is necessarily networked. The distance between the soldier and the gamer is thus further blurred. As Tumber and Webster (2006: 33–34) stress, game theory and digital simulations are essential elements in the conduct of contemporary warfare. While David Nieborg (2006) notes that the same military simulations are used by both soldiers and gamers, Edward Castronova's analysis (2005: 234) goes somewhat further when he suggests that 'the emergence of open-source military game-building tools has effectively turned the entire world into a giant military research lab.'

King and Krzywinska (2006: 199) suggest that gameplay may be used to train players in the techniques of realworld warfare. This facility is not only, of course, the province of the United States and its political, military and ideological allies. Such video games as *Under Ash* (2001), *Special Force* (2003) and *Under Siege* (2007) have promoted (and have been used to train) anti-Israeli paramilitary groups in the Middle East; while even such overwhelmingly neoconservative toys as *Counter-Strike* (1999) might also offer, as Castronova (2005: 231) suggests, a convenient tool for training terrorists.

This training is as ideological as it is strategic. The First Person Shooter game foregrounds the gun, both visually and linguistically. The first person (the ego) becomes identical to the 'shooter' – both the person who shoots and the gun itself, and indeed also the imagined lens of visualization. Within that one word – *shooter* – the distinction between subject and weapon dissolves. The player is translated into an organ of war shooting forth its deadly seed to inseminate a new world order. *America's Army* has promoted itself beneath the slogan 'empower yourself – defend freedom'; yet what it offers is disempowerment, a loss of freedom, a loss of self and even the player's actual death (it recruits you; it can get you killed). It therefore seems significant that, as Lars Konzack (2009: 39) points out, the multi-player mode of *America's Army* deploys an extra level of illusion in order to sustain its players' ideological assimilation: each player sees themselves as the U. S. soldier and the other as the terrorist – though each is always also the terrorist from the other's perspective.

The illusion of agency

Those familiar with the video-sharing website YouTube may be aware of the practice by which digital gamers have edited together clips of – or merely recorded extensive sequences of – their gameplay, added music or voiceovers and credits to it, and posted it online as a so-called *movie*. The experience of watching one of these gameplay films is nauseating: not only in a physical sense (like being a passenger in a drunk driver's car) but also in existential terms. This is the Sartrean nausea which accompanies the realization of one's own lack of control, the revelation that one's feeling of self-determination was only ever an illusion – that the experience of playing the game and of watching the game being played are, in the end, the same. The amateur gameplay video therein exposes the possibility that the digital game's defining sense of a player's agency may be illusory, the realization that the illusion which re-envisages the immutable, impersonal edifice of the game as an extension of the gamer's own subjectivity – as if the player could somehow reconfigure the programmatic structure of the game, could escape its pre-programmed linearity (which is a multilinearity, but is still a linearity, and a finite one at that) – is the precise opposite of what is in effect taking place. In watching gameplay with the detachment of an uninvolved audience, the game's inescapable parameters and its pre-programming become visible.

Lars Schmeink (2008) has referred to the mere *sense* of agency afforded by the video game; for, as Tanya Krzywinska (2008) has suggested, 'you are promised some kind of agency, but your agency is taken away from you.' This sense of agency is always, she adds, a fictive agency. James Newman (2002) has argued that videogames are not interactive and Dominic Arsenault and Bernard Perron (2009: 119–120) have similarly challenged the popular notion that the video game is a predominantly interactive medium. They argue that in fact players are not active but reactive – that players respond to pre-programmed structures within the game, structures designed to predict and react to the gamers' responses. The illusion of interactivity sponsors a sense of agency – but this agency has been externally predetermined or pre-designed.

This illusion of agency is central to the pleasure of the video game. Klimmt and Hartmann (2006: 138) write:

> Most games allow players to modify the game world substantially through only a few inputs […] players often need only a few mouseclicks to fire a powerful weapon and cause spectacular destruction. The ability to cause such significant change in the game environment supports the perception of effectance, as players regard themselves as the most important […] causal agent in the environment.

The game sponsors an illusion of agency which places the player at the centre of its universe – which thereby for the player becomes *the* universe. This illusion is not immediately recognisable as such; from the player's perspective the game's devices remain invisible.

The game's self-proclaimed interactivity is not a case of co-authorship: the gamer is funnelled through a limited and limiting series of preset positions. The digital game constructs and delineates its citizen-user-consumers as avatars of its own ideologies. The gaming subject is *interpellated*, in Marxist theorist Louis Althusser's sense – is hailed or recognized as the central character in their sphere of existence – and, as this process is never without an ideological destination, the subject is posited within (or, in the case of *America's Army*, literally *recruited* to) a new reality. 'Ideology,' writes Althusser (2006: 118), 'recruits subjects among the individuals (it recruits them all), or transforms individuals into subjects (it transforms them all).'

This represents a process of ontological and ideological transformation. To what extent then does the gamer become subsumed to the subjectivity of the game? Although Bolter and Grusin (2000: 253) have suggested that the visitor to virtual reality remains aware of the differences between the virtual and material worlds, they have nevertheless supposed that virtual reality changes our notions of selfhood. More recently King and Krzywinska (2006: 198) have asked:

Are players, really, interpellated to any significant extent into the *particular kinds* of subjectivities offered by the in-game diegetic universe? [...] Plenty of markers exist that clearly announce the large gulf that exists between playing a game [...] and engaging in anything like the equivalent action in the real world. But there are, also, certain homologies. How far these come into play depends on a number of factors, including the [...] forms of realism [...] which can shape the extent to which the game experience approximates that of the real world.

Yet, as has already been suggested, the extent to which the world of the game approximates a prevalent notion of the real world may matter rather less than the degree to which the material world resembles the gameworld. As the reality of the game becomes the dominant mode of being, King and Krzywinska's gamer is increasingly assimilated within the gameworld's subjectivity.

Boellstorff (2008: 120) has observed the formation of distinct identities in virtual worlds. One might argue that the senses of safety and of empowerment generated by the virtual environment give the user an impression of autonomy that fails to register – and therefore to resist – the constraints and influences placed upon it, and that therefore the blurring of offline and online identities through what Boellstorff (2008: 121) calls a 'permeable border' between selves might foster modes of offline being inconsistent with the traditional paradigms of the material world. Boellstorff (2008: 122) comments that a number of the subjects of his study 'spoke of their virtual-world self as "closer" to their "real" self than their actual-world self.' If it is the case, as this suggests, that the virtual self becomes the user's primary benchmark of subjectivity, there might appear to be something increasingly problematic in this identity seepage between these two modes of being – when, as Evans (2011: 34) suggests, 'living within virtual worlds is changing notions of who we are.'

Edward Castronova (2005: 45) has proposed that the gaming avatar is no more than an extension of the player's body into a new kind of space – as though the assumption of a mask or a persona does not transform one's identity. Slavoj Žižek (2008: 83) adopts a rather more ontologically problematic perspective: 'when I construct a false image of myself which stands for me in a virtual community in which I participate [...] the emotions I feel and feign as part of my onscreen persona are not simply false. Although what I experience as my true self does not feel them, they are none the less in a sense true.' Gameplay constructs an alternative but real subjectivity, and, insofar as the gamer increasingly experiences the virtual world as her primary reality, then that alternative subjectivity may come to represent the player's dominant sense of self. Indeed, as Žižek (1999) has also suggested, the game both constructs and constrains the self – and therefore this play is far from liberatingly transformative: 'the much celebrated playing with multiple, shifting personas [...] tends to obfuscate (and thus falsely liberate us from) the constraints of social space in which our existence is caught.'

David Myers (2009: 48) suggests that 'when we play with self, that self is something other than what it is: an *anti*-self' – an alternative, and possibly overpowering, mode of being. Ian Bogost (2006: 136) also recognizes the tensions between these subjectivities – at the point at which the reality of the game blurs with the material world. Sébastien Genvo (2009: 135) has suggested that the video game player may be 'engrossed in his game although he knows that after all it is only a game.' The notion of the integrity of identity in the face of cultural or virtual immersion requires, however, the existence of an *a priori* subjectivity – founded upon the romantic notion of an essence of selfhood – or upon the prioritization of material experience as somehow more influential upon the propagation of subjectivity than digitally mediated experience (as though our physical interactions might for some reason mould our identities more forcefully than those hours spent in the virtual space of the electronic media).

There is, of course, no difference between material and virtual experience: it is just that we tend to use the word 'virtual' in depicting forms of experience mediated by more recently evolved technologies. We are defined by performance and play as much as by 'real life' activity – insofar as there is, of course, no difference between these phenomena, except one imposed by culturally, economically and ergonomically determined epistemologies. Jean-Paul Sartre (1969: 59) famously described the way that a waiter in a café plays at being a waiter: 'all his behaviour seems to us a game.' The waiter is playing at being a waiter: he is playing at being himself. Sartre's point is that it is such play which defines identity: as existence precedes essence, the parts we play define our subjectivities. If we are all playing parts, then the parts we self-consciously play cannot be differentiated from those roles we unconsciously assume. We are all method actors, and, as Camus (1975: 75) suggests, the roles we create and perform return to create and perform us:

> To what degree the actor benefits from the characters is hard to say [...] They accompany the actor, who cannot very readily separate himself from what he has been. Occasionally when reaching for his glass he resumes Hamlet's gesture of raising his cup. No, the distance separating him from the creatures into whom he infuses life is not so great.

The ancient Chinese writer Zhuangzi famously imagined that the man who has dreamt himself a butterfly does not know if he is not a butterfly who is now dreaming himself a man. It is clearly unclear which of our performed or imagined selves is the real one, or indeed whether there is any such thing as a real one. When immersed in performative activity (as we always are) our suspension of disbelief creates an identity for whom our belief is permanent and absolute.

In his crime novel *Virtually Dead*, Peter May describes the attractions and seductions of the virtual world of Second Life. May himself spent a year researching Second Life, even going so far as to establish his own virtual detective agency therein. The central character of May's novel, Michael Kaplinsky, takes on the persona of crime-fighting Chas Chesnokov in Second Life, a role which impacts upon his own identity, existence and survival in the material world (2010: 105): 'Who would he be when he logged out again? Michael or Chas? Or was it possible that, with time, more and more of Chas would return with him to RL?' Indeed Second Life eventually comes to seem to him preferable to real life (May 2010: 142): 'If only he could just be subsumed to the virtual.' He discovers that 'in a world where the reality is virtual, and completely unreal, it is far easier for us to be our real selves' (May 2010: 193). Yet this real self – this increasingly dominant self – is clearly different (although decreasingly distinct) from the original material subject. The protagonist of Andrew Blackman's novel *A Virtual Love* (2013: 53–56) similarly argues that his online profile is his identity – it is not that it is 'not real' – it is 'not fake' – but is the identity which he chooses 'to show the world' in that he maintains 'different identities for different places' – equally real though each of these is.

But if there is no difference between the ways in which material and digital experience construct subjectivity, should the notion of identity within the virtual realm in any way concern us? What is different, of course, about contemporary digital culture is its increasingly globally homogeneous nature, and (through the speed and seamlessness of its operation) the ease with which it disguises its ideological and economic construction. The virtual environment, like any mode of conventional realism, smoothes out the wrinkles in material reality, offering a realm whose continuity of logic makes more sense (and appears more realistic) than the incoherence of the material world. Its realism offers an immersion in the ultimate escapist fantasy – the fantasy of ontological logic, the fantasy that the reality of experience might eventually *make sense*.

John Fiske (1987: 24) has suggested that conventional realism ensures 'that all links and relationships between its elements are clear and logical, that the narrative follows the basic laws of cause and effect, and that every element is there for the purpose of helping to make sense.' Material reality of course lacks this seamless continuity: a cosy continuity which makes things so understandable that we do not make the effort to understand them. The video game, however, embraces its user in that reassuring logic.

Even if she were not lulled into critical complacency by the faultless logic of the virtual experience, it is also apparent that the digital game's speed of operation barely allows its user time for such independent reflection. In their discussion of early film, Adorno and Horkheimer (1979: 127) argue that the relentless speed of the cinema prevents the possibility of reflective thought on the part of its audience. The velocity of the virtual world of the early twenty-first century of course leaves Adorno and Horkheimer's cinema standing.

The transcultural uniformity and universality of the digital domain also suggest a paradigm shift in the mediation of identity. For the first time in human history, the cultural difference which gave that history its momentum appears to be in the process of being replaced by a single world view, a ubiquitous and monolithic mode of mediation, representation and perception. It is not the 'virtuality' of digital culture so much as its globalization which underpins its potential to determine subjectivity.

The homogeneity, seamlessnness, rationality and apparent safeness of the virtual environment are precisely the factors which may reduce its users' ability to resist its influences. When we are immersed into the subjectivity of our chosen avatar, we tend not to notice the extent to which that avatar may have chosen us (insofar as our selection is anticipated and determined by the avatar's own design), and the impact that it may thereby have upon us. (Cf. Narain 2014.)

But the most powerful hold that the virtual environment maintains over its users must surely lie in the impression of empowerment which it offers, an empowerment founded upon the promise of interactivity.

The illusion of interactivity

A still influential view of textual interactivity was advanced some decades ago by the cultural theorist Roland Barthes in his elucidation of the *scriptible* text in *S/Z* and in his celebration of 'The Death of the Author'. Barthes argued that traditional texts represent what he called the *lisible* or readerly: fixed, final and finite products, rather than ongoing processes of interpretative production (Barthes 1974: 5). Barthes's antithesis to this classic readerly text is the writerly or *scriptible* text. Barthes's textual ideal is founded upon the premiss that the progressive function of literature is to transform the reader from a passive consumer into an active producer of meaning (Barthes 1974: 4). The writerly text invites, embodies and requires cooperation and co-authorship: it understands that meaning is an act of interpretation rather than of intention or expression. As Barthes (1977: 148) proposes, the intertextual polysemy of the work of art originates where it is destined to end: in the mind not of its author but of its audience.

In contrast to Roland Barthes's textual idealism, there is a school of critical thought which suggests that the political function of popular culture is to dumb us down and that new technologies intensify this process. Noam Chomsky (1989: 14), for example, has famously argued that 'the media are vigilant guardians protecting privilege from the threat of public understanding and participation.' Adorno and Horkheimer (1986: 120–167) had similarly complained that cinema's homogeneous processes divested its audiences of the power of critical thought. Bertolt Brecht (1978: 187) meanwhile imagined the users of industrial culture as ideological zombies. Brecht's image all too easily fits the stereotype of the TV addict or the video game junkie. Yet, rather more recently, the likes of John Fiske and Stuart Hall have argued against the absolutism of these hypodermic theories of mass-cultural influence: 'I do not believe that "the people" are cultural dopes; they are not a passive, helpless mass incapable of discrimination and thus at the economic, cultural and political mercy of the barons of the industry' (Fiske 1987: 309). Roland Barthes's *scriptibilité* anticipates Stuart Hall's notion that the act of decoding a text may not be equivalent to the process of its encoding – but may encompass negotiation with, or opposition to, the dominant meanings privileged by the position of authorship.

Yet perhaps no texts or media forms are truly negotiable or interactive in themselves. Rather than Hall, Barthes or Fiske's celebrations of the potential for audience co-authorship, it may be that Walter Benjamin's ambivalence offers the most convincing theoretical position. Benjamin (1992: 232) proposes that 'a man who concentrates before a work of art is absorbed by it [but] the distracted mass absorb the work of art.' The former state of immersion permits the survival of an integral subjectivity (not a pre-existent essence but a pre-textual self); the latter process incubates an ideological identity within the passive subject. We remain caught between these positions – between the liberal's free-thinking citizen and the Marxists' dope – or perhaps, rather, we are both (and neither) of these at the same time. We can only be the former when we believe we are the latter; when we believe we are the former, we become the latter.

If those popular texts, technologies and practices which invite audience participation (reality television, competitions and lotteries, phone-ins, teleshopping, electronic government, citizen journalism, Facebook, Twitter, Wikipedia and YouTube, online gambling and digital games) in fact offer only an *illusion* of interactivity, then – rather than promoting participation – they may in fact serve entrenched structures of power by sublimating our desires for active, participatory citizenship. It may be argued that the video game's illusion of *scriptibilité* seduces the player into neglecting the modes of critical negotiation which might prevent the states of ideological assimilation envisaged by Adorno and Brecht – that the game's demands for functional reactivity promote an illusion of agency which lulls the player into an interpretative passivity, and which thereby serves to posit its subject within a virtually invisible (and therefore virtually irresistible) ideological mould. This illusion is central to any process of textual interpellation, but the digital game reinforces it with an apparently unprecedented degree of influence. The video game is neither more nor less interactive than any other mode of textuality – yet the video game announces its interactivity more forcefully than perhaps any other media form.

We might therefore add a third category to Roland Barthes's classification of *scriptible* and *lisible* texts: the *faux-scriptible* text which proclaims its openness to interactivity, which gives its user the illusion of meaning, power and active participation, and which, in appearing to satisfy its audience's desire for agency, in fact subsumes and dilutes that desire. This process resembles a kind of textual karaoke: its audiences believe that their participation represents a form of activity, a mode of agency, but they are, in effect (and in consequence), mere puppets of the text. This *faux-scriptible* text is thus significantly more reactionary and compelling than the *lisible*.

Lev Manovich has argued that this 'myth of interactivity' offers its audience only an illusion of agency. Manovich (2001: 61) writes:

> Interactive media asks us to click on a highlighted sentence to go to another sentence. In short, we are asked to follow pre-programmed, objectively existing associations. Put differently, in what can be read as a new updated version of French philosopher Louis Althusser's concept of interpellation, we are asked to mistake the structure of somebody's else mind for our own.

It is not that the digital game is necessarily any less interactive than any other text; it is that its claims to greater levels of interactivity (and therefore of audience agency) are seductively misleading.

Let's stop playing games

New media have, in their proliferation and domination, attempted to erase the traces of their own material production. In doing so, they refute the possibilities of metatextual reflection which might afford their audiences a Barthesian writerliness. Bolter and Grusin (1999: 24) have written of digital technology's attempts to deny its own mediated character through the promotion of an invisible interface. Our immersion in the digital experience allows us to ignore the fact that the experience is merely digital; this denial remains essential to the processes of immersion.

Ernest Adams (2009) has suggested that video games are like Victorian novels: despite their claims of interactivity, they remain, in narratological terms, classic realist constructs which eschew the disruptive and liberating possibilities of a metatextual *scriptibilité*. The enduring presence of the fourth wall sequesters the player against a self-consciousness which might foster a reassertion of the critical self. The mass-market digital game's refusal to bare its aesthetic devices thereby allows its ideological mechanisms to go unchallenged and unseen. Though the popular video game may permit a tacit awareness of its artificiality (it may let you know it cannot really kill you – not at least straight away), it cannot afford to admit to any challenge to the coherence of its artifice, because that would be to undermine its interpellating structure of realism. Its conventions – its rules – can be asserted but cannot be questioned. Playing a commercial video game is therefore rather more like walking – or boarding a rollercoaster or an aeroplane – than it is like reading *Finnegans Wake*: the more one thinks about it (the more one questions the natural cogency of its conventions of function and production), the less possible it becomes.

There remain, of course, hopes that the digital game, a media form still in its infancy, will eventually adopt modes of greater complexity and ambiguity, of *scriptibilité*. It may be that such games as independent designer Gonzalo Frasca's *September 12* (2003) and *Madrid* (2004) suggest a route towards the textual sophistication of the digital game. *September 12* argues and enacts the futility of the War on Terror (you play an American bomber pilot above a Middle Eastern town, but the more enemies you kill the more they multiply); it is a game which the player can only win by refusing to play. 'This is not a game,' its instructions announce: 'This is a simulation. This is a simple model you can use to explore some aspects of the war on terror.' Frasca's *Madrid* – a sombre pause for reflection upon the 2004 train bombings in Madrid – offers one static graphic (an image of people holding candles) and has only one function and one rule: 'Click on the candles and make them shine as bright as you can.' These arthouse games' modes of interactivity – their incitements to interpretation – do not rely on conventions of realism in their graphic representations, but take place at the level of arguments which they do not articulate so much as they allow the player to explore and perform. These games, like Barthes's *scriptible* texts or Benjamin's optimal artistic experiences, do not permeate and assimilate their players – instead, their players navigate, negotiate, translate and reconstruct the meanings offered by the games.

Gonzalo Frasca (2007: 133) has pointed out that in *September 12* 'the player soon realizes that the game cannot be won [...] the game sabotages itself [...] it does not provide a solution to the problem of terrorism but it aims to show that following the only allowed strategy (shooting missiles) is doomed to failure.' Frasca (2007: 17, 28) argues that while games may be seen as transmitting ideologies some (including his own) are explicit in displaying their politics. This overtness diminishes the possibilities of subliminal ideological indoctrination.

Frasca (2007: 53) argues that games do not require their players' agency so much as the appearance of that agency: 'it is enough for them to believe that they are in control.' A game, says Frasca (2007: 70) is something in which the players believe they enjoy a mode of active participation. Frasca (2007: 89) notes that, while games may appear to offer agency and liberation, 'play may connote freedom but it is always constrained and, therefore, authored – either by a designer but also by the environment and cultural conventions.' Thus, Frasca (2007: 140) argues that games' design techniques guide and privilege certain performances over others, and suggests that in this way games deploy persuasive strategies to interpellate their users.

For Frasca (2007: 201) there remains a constant balance between the player's apparent autonomy and the game's regulatory structure: although the game can generate a vast array of meanings, that range is neither infinite nor extrinsic to the rules of play. One is struck by the parallels between Frasca's definition of gameplay as rhetoric and the post-structuralist philosopher Jacques Derrida's view of textual meaning as a game, one which remains within certain prescribed limits. Within this context, Roland Barthes's notion of the *scriptible* text or textual interpretation (of the textuality which invites co-authorship, or indeed of the textual approach which refutes the primacy of authorial authority and which imposes such co-authorship upon the text) seems apposite. There are limits to the extent to which we can rewrite the rules of Frasca's and Derrida's semantic game-play, but insofar as we are critically aware of those limits – as we surely must be when we play *September 12* (in that the game draws our attention to its own problems, paradoxes and limitations) – then we can become critical and creative players or interpreters of the game/text.

But to what extent, we may ask, is this reflective co-authorship possible when the player is engaged in a commercial, mass-market digital game – a fast-moving First Person Shooter, say? While this approach is clearly possible, it is also reasonably improbable. How might we establish grounds for popular negotiations with populist games which will not acknowledge the need for such empowerment (insofar as they purport to have already empowered their players by putting virtual guns into their hands)?

Frasca has argued that video games 'require a lot of work from the player – much more work than most other media' and suggested that it might therefore be possible for mass-market games one day to foster a degree of critical agency: 'It depends on the game. But I think that they could, within the right parameters. In a classroom setting, with the guidance of an educator, they can definitively do so. By themselves, it's a tougher challenge.' He has added however that he has 'no idea' whether games that encourage critical reflection could ever be popular in a mass commercial market: 'I wish they were but there's no way for me to know this. The situation can change eventually.'

In 2006 al-Qaeda's Global Islamic Media Front published *Quest for Bush*, an anti-American modification of Jesse Petrilla and Bob Robinson's propagandistic and xenophobic *Quest for Saddam* (2003). Two years later the Iraqi-American artist Wafaa Bilal produced *The Night of Bush Capturing: A Virtual Jihadi* (2008) – his own deconstructive reinterpretation of al-Qaeda's modification of Petrilla and Robinson's original game. There is both political and artistic scope within this hijacking of commercial game formats to sponsor new processes of meaning; yet such radical and subversive games continue to languish in commercial obscurity. When in 2009 a Scottish games company collaborated with former Guantanamo Bay inmate Moazzam Begg to produce *Rendition: Guantanamo*, it was met with public outrage and its launch was cancelled.

The digital game's illusion of interactivity disguises its indigenous culture's innate conservatism. The commercial video games industry has put an extraordinary emphasis upon the significance of interactivity in its field of production; but while the digital game may be neither more nor less interactive than any other text, this emphasis itself seems not only to mislead its consumers but also to disempower them. In a TED talk of February 2010 Jane McGonigal proposed that 'games like World of Warcraft give players the means to save worlds, and incentives to learn the habits of heroes' and asked 'if we could harness this gamer power to solve real-world problems.' This is, of course, exactly what *America's Army* has attempted to do. When, then, McGonigal's dreams of a world of *World of Warcraft* are eventually realized, this new reality may not represent the democratic cybertopia of which some have dreamt. W. H. Auden (1979: 81) suggested that 'each in the cell of himself is almost convinced of his freedom' – yet when each individual is not *almost* but *absolutely* convinced by that illusion of self-determination, those who are interpellated by the dictatorship of the algorithm will not even dream of autonomy and liberation, because (like all those sustained by the *faux-scriptible*, like the victims of *The Matrix* itself) they will mistakenly believe that they are already the authors of their own destinies.

Transfigurations

The protagonist of James Cameron's 2009 film *Avatar* eventually discovers himself becoming his avatar: 'everything is backwards now – like out there is the true world and in here is the dream. I can barely remember my old life. I don't know who I am anymore.' Thus Cameron's hero chooses to surrender his old life and his old body in favour of the virtual and fantastical world in which he has found himself – or, rather, in which he has reinvented himself – or, rather, in which his self has been reinvented.

Cameron's hero's subjectivity becomes at once liberated by and subsumed to the existential condition of his avatar. What Dixon (2007: 241) has called the 'digital double' is always, eventually, an uncanny double, a doppelgänger which, as Freud (1985: 335–376) suggested, returns to question the integrity of the original self. The projection of the self comes to expose the subject's lack of autonomy, insofar as the projection (a product of mass mediation) at once overwhelms the original subject and reveals the fact that this subject has never itself been anything more than a similar product of such mass mediation.

The *Oxford English Dictionary* defines an avatar as 'the descent of a deity to the earth in an incarnate form.' Yet, while Cameron's futuristic film might finally allow its protagonist the physical possibility of that apotheosis, the virtual avatar of contemporary media affords its user only the illusion of that divine power – whether as a player of digital games or as the viewer immersed in James Cameron's 3-D world. This is not the incarnation of a god in human form but the illusory manifestation of a human in the image of a god. The pseudo-utopia of the cinema, the video game or the internet does not as such represent a process of empowerment but one of consolation.

The *Oxford English Dictionary*'s definition of the word 'agent' is also perhaps worth considering in this context: 'one who [...] acts or exerts power [...] one who acts for another, a deputy [...] the material [...] instrumentality whereby effects are produced; but implying a rational employer or contriver.' When the subject considers herself an agent then she may identify herself as one displaying agency or autonomy; and yet an agent is also a functionary or operative of an agency or authority. The hero of James Cameron's *Avatar* transcends his status as an operative of military-industrial authority to achieve some level of autonomy (he will not serve) – but this is of course a matter of fantasy, a wish-fulfilment which interpellates subjectivity and sustains ideological subjection. If we are all *secret agents*, then the defining secret (the secret in which we are complicit in hiding from ourselves) is that we are agents in the subservient rather than the autonomous sense of the word. That is not to suggest that there is a rational contriver (any more than it is to require an intelligent designer in order to explain evolution). As Foucault (1991) and Bourdieu (1977, 1986) have supposed, such self-perpetuating structurations seem more the products of unconscious complicity than of conscious conspiracy.

Bourdieu (1977: 85) depicts the process of mass culture as a work of inculcation by which history's objective structures perpetuate themselves through human individuals. This process, for Bourdieu (1977: 79), involves a mediated illusion which allows the individual human agent's actions to appear to be the product of reason. This consoling notion of the ministration of conscious and reasonable agency (of the individual subject, of an autocratic elite, or of a divine architect) – what Bourdieu (1991, 2005) has called the *mystery of ministry* – is only as comforting as it is successfully illusory. Despite all the evidence, rational agency remains an idea to which we relentlessly cling.

An experiment conducted by Soon et al. (2008: 543) has demonstrated that 'the outcome of a decision can be encoded in brain activity up to 10 seconds before it enters awareness.' These neuroscientists hooked up a test subject to a brain scanner and asked her/him to press one of two buttons: the subject was asked to press the button as soon as s/he had decided which button to press. As early as ten seconds before the subject pressed the button, the scientists were able accurately to predict (based on which areas of the brain were shown by the scanner to be active) which button the subject would press. The implications of this experiment are somewhat disturbing: that the brain makes decisions before the conscious mind becomes aware of those decisions, and that therefore one's consciousness, one's subjectivity, is not in control: subjective agency (or 'free will') is an illusion.

We might go further and envisage an extension of this experiment: imagine that we tell the test subject that they should themselves choose which one of the two buttons to press unless (before pressing the button) they receive an instruction from the scientist as to which button to press (they should then press the button as instructed); the scientist then tells them to press the button their brain has already decided upon (although this decision has not yet filtered through to their consciousness); they therefore press the button in accordance with their own brain's decision, but in the belief that they are doing so in response to the scientist's external agency. This exposure of the deterministic nature of being advances an existential alienation whose disillusionment may be the closest thing to empowerment that we can hope to achieve – insofar as our only empowerment can be the recognition of the irrevocability of our disempowerment.

Yet what *seems* to us to be most empowering is, for the most part, the very antithesis of this: not the recognition of our disempowerment but the comforting illusion of our empowerment. This illusion of effective agency is something which digital games, or for that matter reality TV shows, offer us in abundance.

CHAPTER 4

Reality Television

Reality television offers interactivity to its audiences: the opportunity to immerse themselves in the 'real' social interactions of 'real' people, and very often also the experience of direct interaction through voting for their favourite or least favourite contestants. In its most popular form (in such series as *Big Brother* and *The X Factor*) reality television has become an ostentatiously interactive multimedia experience, courting audience participation online, by telephone and through the use of the interactive functions of digital broadcasting. This chapter will however contend that this sense of interactivity may be as illusory as the reality which these programmes purport to reflect.

One of the most insidious and exploitative aspects of purportedly interactive mass culture is, as Papacharissi (2010: 65) suggests, the way in which 'the guise of participatory media and the promise of power' entice audiences to produce content without ever being compensated for their work. Yet there is an illusion of compensation: the magnanimity of the process by which the medium offers to elevate and empower its participants, the gift of symbolic capital which it confers upon its subjects and (both vicariously and by offering a simulacrum of democratic or creative agency) upon its audiences. But, as Pierre Bourdieu (1977: 195) has argued, this gift of symbolic agency is one which, insofar as it claims to be both free and liberating, silently binds its subject within its established power relations: 'a gift which is not matched by a counter-gift creates a lasting bond, restricting the debtor's freedom and forcing him to adopt a peaceful, co-operative, prudent attitude.' This process of ideological assimilation and hierarchical entrenchment necessarily disguises itself beneath a veil of liberation. Oppression masked as liberation through an illusion of agency affords and manufactures the consensus essential for the reinforcement and perpetuation of power structures – inasmuch as, again in the words of Pierre Bourdieu (1977: 195), 'the endless reconversion of economic capital into symbolic capital [...] which is the condition for the permanence of domination, cannot succeed without the complicity of the whole group.'

This phenomenon is seen clearly in the proliferation of reality television; through, for example, the divine generosity of Big Brother or Simon Cowell. Tincknell and Raghuram (2004: 263) have argued that 'the *idea* of agency' was crucial to the success of the ground-breaking series *Big Brother* – but that this should not imply that the audience was granted any real agency or textual autonomy:

> the extent to which this constituted the power to determine the meaning produced *in* and *by* the text is debatable. The production company [...] certainly emphasised the importance of audience participation [but] clearly retained editorial control of what was seen and heard on the programme.

The new reality

As Geoff King (2005: 94) proposes, the paradox of television is that it has come to seem at its most realistic when it is in fact at its most mediated. Jean Baudrillard (2005: 75) suggests that in the simulacrum of reality television we are witnessing 'the confusion of existence and its double' – a confusion between reality and representation, and a prioritization of the representation of material existence over its reality – a confusion then, between reality and television.

The banalities of reality television have come to assume a historical significance. In July 2010 Australia chose to reschedule a televised election debate from its traditional slot in order to avoid a clash with the reality show *MasterChef*. The political has been subsumed to the demands of the mass entertainment industry. Australia's version of *MasterChef* has developed sufficient public significance that in July 2011 the programme managed to attract the Dalai Lama to act as a guest judge. On 19 July 2011 the *Daily Mirror* reported that 'stunned contestants prepared lunch for the Tibetan spiritual leader – but he refused to rate their offerings, saying it would be against his Buddhist principles.' Meanwhile the British TV chefs Jamie Oliver and Hugh Fearnley-Whittingstall have become known for their food-related political campaigns; and in spring 2012 it was revealed that fans of Egypt's popular TV cook Ghalia Mahmoud had urged her to stand for the nation's presidency. That September British TV cook Clarissa Dickson Wright also hit the headlines when in relation to a controversial plan to cull the nation's badgers she suggested that we should eat the beasts.

Reality television has exacerbated an ongoing confusion between politics and celebrity – as between news and entertainment. In February 2013 it was announced that Nelson Mandela's granddaughters were (with their grandfather's blessing) set to star in a reality television programme called *Being Mandela*. Two months later, however, the *Sowetan* reported that the series had met a scathing critical response from South Africans in what it described as 'a baptism of fire on social media.' February 2013 had also seen the killing of South African model Reeva Steenkamp by her partner, the athlete Oscar Pistorius. Her death did not however prevent a South African TV channel from screening *Tropika Island of Treasure* – a reality television series featuring Ms Steenkamp – that same month. In March 2014 the media exploitation of Pistorius's subsequent trial outraged many, including the presenters and viewers of Channel 4's topical comedy talk show *The Last Leg*: first, when a British bookmakers started offering odds on the outcome of the trial – and promising punters their money back if Pistrorius (a double amputee) were to 'walk'; then, when one viewer pointed out that Sky's 'Pick TV has the Oscar Pistorius trial listed as entertainment.'

In May 2009 Piers Morgan, a former editor of the *Daily Mirror*, a future former CNN talk show host and a judge on *Britain's Got Talent*, had described one of that programme's contestants (a Scotswoman whose unprepossessing appearance had garnered her international fame) as an antidote to global recession. In doing so, Morgan invoked a typically media-centric perspective upon contemporary society, one which promotes the mass-mediated image above material substance and which (in the era of the spin doctor and of reality television) seems virtually unassailable. In

April 2014 it was widely misreported that Morgan had been appointed as a media consultant to the UK's Liberal Democrats.

As Biressi and Nunn (2005: 144) have reminded us, the proliferation of reality television has coincided with a period in which politics has been increasingly packaged as media product – and it seems that just as politics has blurred into the field of popular entertainment, so entertainment has assumed a new socio-political significance. Indeed, Ouellette (2009: 240) has gone so far as to suggest that reality television may be seen, in some of its manifestations, as an active agent in the neoliberal transformation of democracy.

House of cards

Maitles and Gilchrist (2005) have pointed out that more people voted for the winner of Channel 4's *Big Brother* than the combined vote for the Scottish Parliament, Welsh Assembly and London Mayoral elections between 1999 and 2000. Kilborn (2003: 15) meanwhile asks what it tells us about our culture that more people vote on *Big Brother* than in European elections. As the comedian Dave Gorman pointed out in his TV series *Modern Life Is Goodish* (2014), 'the media is obsessed with us having our say. We've never had more say. We've also never voted less.'

This late postmodern era appears to be witnessing a process whereby material history is increasingly subsumed to a mediated virtuality, to the reality of television, of reality television, a transformation of the material and the substantive into a mass media product. This much has been painfully apparent since (in January 2006) British MP George Galloway appeared on *Celebrity Big Brother* in a leotard and pretended to be a cat – or since (in January 2007) racial tensions on the same show prompted the burning of effigies of one of its contestants in the Indian city of Patna and apologies from the UK Chancellor during a visit to India – or since (in July 2007) the Prime Minister was obliged to comment on that franchise's next race row – or since (in March 2009) the Prime Minister led the tributes to the late *Big Brother* star (and alleged racist) Jade Goody.

Stephen Coleman (2006: 457) has helpfully described *Big Brother* as representing 'a counterfactual democratic process in which conspicuous absences in contemporary political culture are played out.' In response to the reduction in political trust provoked by a perceived democratic deficit, Coleman (2006: 477) has gone so far as to suggest that reality television producers might imagine media formats that restore public interest and participation in politics. This perspective appears to be shared by such politicians as Barack Obama and David Cameron who have courted such reality television gurus as Simon Cowell.

Kaur (2007: 11) cites former *Big Brother* contestant (and former Conservative parliamentary candidate) Derek Laud's reasons for appearing on that show: that, while increasing numbers of people were registering their opinions in reality television votes, decreasing numbers were voting in general elections. According to Jonathan Bignell (2005: 96) the British Member of Parliament Jane Griffiths contacted programme-makers Endemol in November 2003 to suggest the launch of a House of Commons version of *Big Brother*. It seems that Ms Griffiths believed that in the hearts and minds of the great British public the political significance, responsibility and privilege of parliamentary democracy had already been subsumed to the pseudo-democratic simulacra of reality television. Jane Griffiths has since added that she 'thought it would make MPs seem more human.' It appears that we need the fantasies of reality TV to make political reality look real.

Channel 4's 2010 series *Tower Block of Commons* – in which members of parliament shared the lives of the residents of British council estates – to some extent explored this format, but, insofar as it promised to focus more upon the experiences of the tower block tenants than upon the personalities of the politicians (who interacted with the tenants but not with each other), it offered to anchor its discourses closer to socio-economic reality than to the depthless representational matrices of reality television. One participant in the series, Liberal Democrat MP Mark Oaten has commented that 'getting politicians away from Westminster, the *Today* programme and party political broadcasts is one way of reaching out to more people. Living in a tower block helped me contact people who never listen to conventional political debate and it taught me a thing or two.'

Another participant, Conservative MP Tim Loughton added: 'If done properly like *Tower Block of Commons* then I think these programmes have the potential to engage more people in the political process who otherwise would not be interested. It also has the potential to challenge common misconceptions about all MPs being out of touch. I received hundreds of emails from complete strangers from around the country saying how the programme had completely changed their view of MPs and Conservative ones in particular.'

However, Labour MP Austin Mitchell, another of the politicians who appeared on the programme wrote on his blog in February 2010 that he regretted his participation in the series, which, he claimed, set out to 'humiliate MPs' – adding that the series was a 'disgrace' and describing those responsible for it as 'bastards'. Mitchell argued that the series fitted the superficial stereotype of reality television by concentrating more upon the characters of the politicians involved than upon the conditions experienced by residents of public housing estates.

Mr Mitchell was also kind enough to comment on his participation in *Tower Block of Commons* for the purposes of this chapter. As not only a reflection on the state of reality television, but also a depiction of the experience of disempowerment involved in participation in the genre, his comments are worth including at length:

> Reality TV is currently fashionable for obvious reasons. It's cheaper than studio-based programmes. The current fashion for reality TV has inundated a population living in a deteriorating reality with escapist programmes.
>
> Reality programmes on social issues can be a major force for raising concerns and exposing a comfortable audience to issues of concern and realities of which they'd otherwise know nothing. Which is the reason why I was upset and annoyed at *Tower Block of Commons*. I thought I was participating in a programme which would show the reality of what it's like living in tower blocks. The producers were actually intending to make fools of MPs by showing that they can't handle the realities of hard lives, which I freely admit I can't. As I see it, they wanted to show MPs as out of touch, high-living figures of fun.
>
> The producers assured me that this programme would be a serious and concerned look at the lives of the people in the blocks. The MPs would be on a voyage

of discovery to see this and project their concerns to a wider audience. This was to be a programme with a mission.

Then the problems began. The director set out conditions which hadn't been mentioned before. These were that I should wear what she described as 'estate uniform' namely T-shirt, tracksuit bottoms and plimsolls. I didn't see any other pensioners in Orchard Park in this kind of outfit and I looked ridiculous in it so I refused. Which upset her and meant lots of questions thereafter from the people they introduced me to on the estate, like why wasn't I dressed like them? Since the other people in the programme could hardly have been bothered about this, I think they were put up to it.

The programme was a series of missed opportunities. I asked regularly why we saw nothing of the other residents in the tower block, why we didn't explore the shops (few and fortified) or the schools, or the lack of facilities for young people. No time for any of that, apparently, but plenty to waste showing me shopping with no idea of prices or me cooking incompetently, changing nappies, and learning how to place bets (none of which are part of my parliamentary duties and none of which I ever do), all totally irrelevant and totally useless so far as the main purpose of the programme, the life of the people, is concerned.

Who cares about me?, I kept asking, but the producers were more intent on putting MPs through humiliations than focussing on the people who have to live in tower blocks.

This doesn't of course degrade the whole category of reality programmes. Honestly done, they can (and should) reveal the conditions and views of a whole range of social groups, let them speak for themselves and put their case to a wider audience. They could bring diverse groups and voices onto the public agenda. That's a democratic advance. It's not real power but it's as close as a lot of people will get.

The speeches of politicians are words spoken into the wind: ephemeral and unheard unless they are illustrated by real problems and messages from real people. Reality TV can do this. I see it as a force for social concern and improvement, but it's also useful if it educates by showing us the views and conditions of other sections of society. Reality TV can bridge the gaps, spread information, illustrate other lives and give a voice to the silent sufferers. That's educative and informative.

However, now that we are in a multi-channel situation, the audience channel-hop around far more than they used to in those days of channel loyalty. This means that producers have to introduce an element of entertainment, or even shock, to get the audience.

Mitchell makes a number of key points: that reality television could clearly function as a powerful instrument in the progressive development of society (it is obviously influential upon society), but that in order to do so it would need to reflect reality rather than manufacture or fabricate the real world. In other words then, it would need to shift its focus from an overwhelming concern with image to a concentration on real material issues. Yet this focus upon image is clearly a phenomenon which influences much contemporary news-making (see, for example, McQueen 1998), and is also central to the practices of modern politics (see also, for example, McLuhan 2001, Greenstein 1967, van Ham 2001 and Savigny and Temple 2010). As Swanson and Mancini (1996: 272) have suggested, contemporary political strategies tend to involve the personalization of telegenic leaders and the sidelining of policies. Or as Jean Baudrillard (2005: 98) has put it, 'the image is more important than what it speaks of.' Within this increasingly virtual culture, the prospects might therefore seem bleak for Mitchell's hope that a return to social realism might restore mass media entertainment's progressive role.

In November 2012 the Conservative MP Nadine Dorries – who had previously also appeared in *Tower Block of Commons* – was temporarily suspended from her party for taking part in the reality TV series *I'm a Celebrity … Get Me Out of Here*. She had, incidentally, been the first competitor voted off the show that season. A year later she apologized to her fellow parliamentarians for failing properly to register income from this reality TV appearance. So much then for the democratizing potential of reality television to bring politics to the people.

The X in the box

In his study of reality television Jonathan Bignell (2005: 130) details a particularly disturbing instance of the confusion between reality television and historical reality – when, on 11 September 2001, the producers of the American version of *Big Brother* called one of the programme's housemates into the diary room to inform her that her cousin who worked in the World Trade Center was among the missing. This confluence of reality television with the major events of material history reminds us that reality television has itself become one of those major events. In an article in *The Sunday*

Times on 3 January 2010 Rod Liddle pointed out that in its review of the first decade of the twenty-first century broadcast a few days earlier, on New Year's Eve 2009, BBC television 'cut from footage of those planes smashing into the Twin Towers to a man called Will Young winning the original incarnation of *The X Factor* – as if these two crimes against humanity were equal in their devastation.' It seems at least that these events are increasingly accorded similar historical significance by the mass media.

In May 2006 *The Observer*'s television critic Andrew Anthony asked: 'Where were you when Shahbaz walked out of the *Big Brother* House? It's not quite the Kennedy assassination, I'll concede, but we can't choose the gravity of the times in which we live.' Reality television has come to replace material history; events are only and essentially media events. In August 2006 Andrew Anthony added that *Big Brother* 'fulfils its Orwellian promise: it *is* society.' Multimedia television is the new society, a post-historical substitute for politics and democracy. When, for example, in January 2010 Tony Blair faced a public enquiry into the Iraq War, *The Daily Telegraph* announced that (in the style of a reality TV show) its online readers would be able to vote live on the veracity of each statement made by the former Prime Minister – with a truth meter on the side of the screen displaying (like a TV talent show's vote counter) the audience reaction to Blair's testimony: 'You'll be able to follow a live stream of the testimony as it happens and give us your opinion by voting on our live lie detector, which follows votes to track whether the viewers believe him or not on any particular issue.' Indeed coverage of the UK's televised election debates a few months later included live audience response meters displaying the ups and downs of the three leaders' popularity ratings as the arguments progressed.

Reality television, like contemporary politics, has transformed issues of substance into matters of image and performance. Reality television does not, as its name might suggest, unintrusively observe ordinary people in ordinary situations: that is the province of fly-on-the-wall documentary. On the contrary it takes extraordinary people (extroverts and exhibitionists, the remarkably talented and the remarkably talentless, eccentrics and celebrities) and puts them into situations to which they are unaccustomed and which to them are extraordinary. The reality of reality television is therefore radically artificial; it derives its illusion of reality

from an aura of immediacy founded at first upon a vicariousness or intimacy of association and later upon a promise of audience participation
or interactivity.

The popular British television series *The X Factor* (2004–) closely fits
this model. It offers members of the public the opportunity at once to
achieve fame and fortune (to win the celebrity lottery), to share the vicarious
pleasure of the achievement of that fame and fortune (through identification
with the contestants) and to influence the contestants' progress (through
a telephone voting system – which also offers its users the chance to win
tickets to join the studio audience, to participate even more closely in the
show – as well as to win a life-changing cash prize). It is, as such, a traditional
talent contest which also affords the modes of pleasure associated with such
archetypes of reality television as *Big Brother*: liveness of performance and
immediacy of reaction, vicarious involvement and a sense of democratic
participation. Yet even this illusion of participation – this mock-democracy,
this democracy of insignificance and ephemera – is limited. Until the show's
final stages the weekly public vote determines two candidates for expulsion
from the process: the programme's judges then decide which contestant to
reprieve and which to discard. It is as if I repeatedly asked you to choose
two cards to reject from a pack, and then just rejected one of these; you
would enjoy the illusion of agency, but we could get through the whole pack
and would still have left the one card I had decided to keep from the start.
The X Factor's celebrity judges in this way continue for the most part to have
the final say: democracy only extends so far; the ultimate power continues
to rest in the hands of a self-appointed elite. There is perhaps a comforting
conservatism in this situation which appears to appeal to audiences alienated from more traditional structures of political trust.

Elizabeth Day (2010: 22) has written that 'part of the attraction is
the sense of control the *X Factor* gives us: the sense that we can put right
wider social wrongs by voting for our favourite contestants' – and that this
illusion allows the programme's audience 'a much-needed sense of agency.'
Day (2010: 20, 22) has further suggested that *The X Factor* affords a sense
of community engagement lost to the denizens of 'a world increasingly
dominated by Facebook and Twitter' – while noting therein the irony that
reality television is itself a similarly virtual construct.

The success and the pleasure of the classic reality television format appears to rely on three key elements: the semblance of reality (the liveness of performance and response, the openness, normalcy and naturalness of performers' personalities, identification with those performers or contestants and the sharing of their heartbreaks and triumphs), the feeling of active participation (through telephone voting to save those contestants you like or to evict those you loathe) and a sense of the significance of the event. The last of these requires the collaboration of other media forms and outlets: *Big Brother* and *The X Factor* become significant events only insofar as tabloid newspapers, celebrity magazines, 'soft' news programmes and entertainment websites recognize and represent them as such. If the first rule of reality television is that you must talk about reality television, then the second rule of reality television is that you *must* talk about reality television. Even Prime Ministers have colluded in this process. In November 2008 *The Daily Telegraph* revealed that Gordon Brown was a closet fan of *The X Factor* after it emerged that he had sent personal letters to the show's contestants. In November 2009 Brown, while still serving as Prime Minister, informed *GQ* magazine that he was an *X Factor* fan – and a few days later told Manchester radio station Key 103 that he was not fond of one particular act on the programme, the teenaged Irish twins John and Edward Grimes (aka 'Jedward'). Brown's opinions were discussed by the programme's judges on the 7 November edition of the series, in which showrunner Simon Cowell asked fellow panellist Louis Walsh to apologize to the Prime Minister for implying that his views on the Grimes twins suggested he was out of touch with the mood of the nation. However, on 22 November 2009 the BBC reported that Gordon Brown had sent his best wishes to the twins in response to negative reactions he had received to his criticism of their performances.

Earlier that month *The Daily Telegraph* had reported that John and Edward Grimes has 'a new celebrity fan' – David Cameron. The *Telegraph*'s use of the term *celebrity* to describe the future Prime Minister was revealing in itself. This confusion of political reality with television celebrity was further emphasized when shortly afterwards the Labour Party's PR machine issued an online poster which showed David Cameron and his then Shadow Chancellor George Osborne morphed into the

Grimes twins – alongside the slogan 'you won't be laughing if they win.' It appeared that loyalty to this series had become a significant test of one's political mettle. Although it is unclear to what if extent (if any) these events impacted upon long-term voter choice, it is perhaps worth noting that the week after the height of this controversy an ICM opinion poll showed that the trend of Conservative gain had been briefly reversed when Labour narrowed the gap between the political parties by four per-centage points.

Each episode of *The X Factor* opens with a glossy title sequence in which a shining 'X' shoots across the solar system towards the Earth, zooming down upon the UK like a falling star. One is reminded in this context of the Christian nativity – or, for that matter, of the opening of Leni Riefenstahl's *Triumph of the Will* (1935) in which the Führer descends from the heavens like a Messiah. These opening titles remind us that this is a pivotal moment in the history of the world: the genesis of an avatar, a divine star.

The reality television show thus announces to its audience that it offers an event which is extraordinary and fantastical and at once both real (historical) and significant (historic): one in which the audience can participate both vicariously and actually. It offers its audience a remark-ably effortless way to assert their subjective significance by participating actively in the historical process – to have the illusion of making history – inasmuch as this is precisely what history now is, insofar as this is how power (media power, which is public power) now views history. The most successful reality television shows offer themselves as historically significant by virtue of their ubiquity, both as global media formats, and as national media events (and the complicity of the popular press is clearly key in this). Hill (2005: 4), for example, notes that half Sweden's population watched the finale of the Swedish version of *Survivor* in 1997, that more Spaniards watched *Big Brother* in 2000 than tuned in to Real Madrid's Champions League semi-final, and that in 2003 Norway's *Pop Idol* pulled 3.3 million SMS votes from a population of 4.3 million people.

The X Factor advances an alternative to democracy, the enigmatic 'X' of celebrity which replaces the ostensibly obsolescent cross on the ballot sheet. It seems no coincidence that when, in March 2010, in the run-up to the UK's 2010 election, BBC television launched a youth version of their flagship political discussion programme *Question Time*, they chose *X Factor* host Dermot O'Leary to front the show. Helen Wilkinson (2010: 47) has proposed that 'we have not yet arrived at the political *X Factor*, though we may not be far off.' Other commentators have suggested it is already here. A month before Britain's 2010 election, in an episode of the BBC situation comedy series *Outnumbered* a small child gave a description of how she imagined the British election process worked: 'Is there like lots of people and then they say "the lines are now open" – and then you vote off all the annoying ones until there's just one left and then they go – "I'm so happy, I'm Prime Minister now"?' Her father responded that she was confusing democracy with *The X Factor*. The implication, of course, is that this is what we are all doing. Indeed in April 2011 *The Independent* published a piece arguing in favour of the Alternative Vote electoral system (in the weeks before a UK referendum failed to introduce such a system) explicitly on the grounds that it mirrored the voting structure of *The X Factor*.

On 14 December 2009 the BBC had reported an idea advanced by *X Factor* creator Simon Cowell to further blur the boundaries between political reality and the escapist fantasy which we call reality television – an idea which, although as yet unrealized, offers a hauntingly possible projection for twenty-first century democracy:

> Cowell has designs on the next UK general election [...] a series of big prime-time shows leading up to the election in which the public would hear two sides of the argument about several issues. There would, he said, be a red telephone for the politicians to ring in, a massive *X Factor*-style studio audience split for and against the issue, and live voting by the viewers.

It was reported that a spokesperson for the Prime Minister had responded to Cowell's idea that 'he welcomed attempts to promote democracy.' (It may be pertinent at this point to recall that in the run-up to the UK's general election in 2005, ITV had launched a reality show called *Vote for Me*, in which the public might select a candidate for parliament. The manifesto of the eventual winner included the mandatory castration of paedophiles and an immigration policy designed to deport 20 million people from the UK.)

In July 2008 *The Times* newspaper reported that Simon Cowell had been brought in by the British Conservative Party to help them 'showcase Tories' talents.' In December 2009 *The Guardian* added that Conservative Party leader David Cameron had suggested that 'there is probably something we can learn in politics' from Simon Cowell. The day before the 2010 election Cowell was afforded *The Sun* newspaper's front page when he formally announced his support for Cameron's Conservative Party – and while he emphasized that he did not believe a general election was equivalent to *The X Factor*, one might therefore wonder why he thought his judgment on the candidates should be so prominently delivered to the readership of the UK's highest selling daily newspaper.

Cowell's support of the Conservative Party was echoed by Andrew Lloyd Webber's endorsement of the Conservatives and Alan Sugar's work on behalf of the Labour administration. In June 2009 it was announced that Sugar would be elevated to the House of Lords and join Gordon Brown's government in the position of 'enterprise tsar'. Sugar, like Lloyd Webber and Cowell, is not only a successful businessman – he also fronts his own reality television series, the BBC's *The Apprentice* (2005–). Because of Lord Sugar's political associations, the BBC chose to delay the screening of the sixth series of *The Apprentice* until after the election of May 2010; on 11 May 2010, however, Sugar expressed his frustration that the reality show featuring Lloyd Webber *Over the Rainbow* had been screened through the election period: 'We saw him in *The Sun* newspaper last week saying he backs Mr Cameron.' Indeed on an edition of *Over the Rainbow* just over a week after the election Lloyd Webber drew parallels between that talent contest and the democratic process: 'each one of these girls has tasted rejection – a bit like our political parties.' While optimists might suggest that this blurring of distinctions between reality television and democratic processes represents a healthy integration of the political within popular culture, others might suppose that this phenomenon results in a dilution of the substantive issues of politics which threatens to transform democracy into the demagoguery of a popularity contest.

The people's choice

The BBC's political correspondent Ben Wright pointed out at the end of April 2010 that the then ongoing election campaign felt like 'a TV election – almost like *The X Factor*.' Two weeks earlier the BBC's political discussion programme *This Week* had opened with a song and dance tribute to *The Wizard of Oz* – an apparent attempt to combine their ongoing election coverage with the TV series *Over the Rainbow*. The programme went on to feature a star of reality television show *Pineapple Dance Studios* offering advice to the UK's three main political party leaders on how to present themselves in the forthcoming prime ministerial debates. Indeed, *The Guardian*'s Charlie Brooker would later describe those televised debates as 'a political version of *The X Factor*.' (Conversely, Brooker has elsewhere depicted the reality television series *America's Next Top Model* as a 'general election for people who'd vote for sequins.')

On 24 April 2010 – at the height of the election campaign – the BBC cited a survey of young children which suggested that many would like to see Simon Cowell elected Prime Minister. The significance of such reality television series to the British political process was underlined by *The Sun* newspaper's front page headline the morning after the election of spring 2010: 'Cameron wins the eXit factor.' *The Sun*'s News International stablemate *The Times* meanwhile led that morning with the headline 'The X factor: David Cameron is close to power, according to exit poll.'

This election had been heralded by many as offering a return of politics to the people. On the day of the election the front page of *The Independent* newspaper had hailed that event as 'the people's election'. It argued that through the country's first ever series of televised debates between the leaders of the main political parties, and as a result of online activities by individual bloggers and posters which challenged the propaganda of the established parties, influence over the political agenda had shifted into the hands of the British people. The results of that election, however, did not demonstrate any significant shift in public opinion away from the two largest political parties; the smaller parties and independent candidates did not make the expected electoral breakthroughs which might have diminished the Labour and Conservative stranglehold over the British political system that had endured for the best part of a century (until a 2014 Euro-election anomaly). Independent bloggers remained for the most part unread, while the world wide web continued to spew forth such democratic delights as the Labour Party's YouTube channel (initiated by Tony Blair in April 2007), a YouTube version of Prime Minister's Questions (introduced by a gurning Gordon Brown in May 2008), and, perhaps most disturbing of all, the video weblog launched by David Cameron in September 2006, Webcameron, on which one could view the great man pontificating on the issues of the day from the comfort of his kitchen.

In May 2010 the polling organisation YouGov reported that 29 per cent of UK viewers believed that the televised debates between the leaders of Britain's three main political parties had been the 'most interesting programmes on TV' over that period. Only ten per cent of those questioned considered the popular reality television series *Britain's Got Talent* the most interesting show then on. Although it is perhaps clear that the heightened interest in the political process may have been more closely related to the closeness of the parties' standing in the polls, it has been suggested by some commentators that the popularity of these televised debates was responsible for the increase in voter turnout at the subsequent election – up to 65.1 per cent from 61.4 per cent at the previous election of 2005.

Figures, however, released by the Broadcasters' Audience Research Board (BARB) do not necessarily demonstrate such a healthy public interest in democracy. The first week of the debates saw *Britain's Got Talent* achieve 11.87 million viewers, and the election debate (screened by terrestrial broadcaster ITV) 9.68 million. Although lagging behind Simon Cowell's reality television show, the debate gained a respectable audience share: this was, after all, the first time such a debate had ever been screened in the UK. The following week, the debate moved to the satellite broadcaster Sky News: while *Britain's Got Talent* took 11.45 million viewers, the party leaders' debate had plummeted to 2.21 million. Back on terrestrial television, the third and final debate in the series – broadcast on BBC One – managed to increase its ratings to 7.43 million viewers. That week, however, *Britain's Got Talent* attracted 11.72 million. Although the first debate had been the second most popular programme of the week (across all British television channels), the third debate was only the sixth most watched programme of the week on BBC One and was also beaten in the BARB ratings by eight programmes on ITV. (The second debate, on Sky News, did not feature in the cross-channel top 50 programmes of that week – indeed BBC One and ITV each managed more than 30 programmes with higher ratings that week.) The cumulative viewing figures for the three debates totalled 19.32 million – giving a weekly average of 6.44 million viewers. Over that three-week period *Britain's Got Talent* averaged 11.68 million. In other words, the average viewing figures for the leaders' debates were just slightly over half (55 per cent) of the ratings for *Britain's Got Talent* over the same period.

Popular entertainment once again trounced national politics. This situation continued in the aftermath of the election itself. On 14 May 2010 the *Daily Mail* newspaper reported that more than a thousand people had complained when (three days earlier) the BBC had chosen to postpone episodes of the soap operas *EastEnders* and *Holby City* in order to broadcast live coverage of the departing Prime Minister Gordon Brown's resignation speech from 10 Downing Street. As the *Mail* pointed out, 'despite historic changes in the real world, it seems many would have preferred to have watched the fictional goings on around Walford's market stalls and in Holby's hospital.'

On 12 May 2010 the BBC observed that an average of 8.9 million people had watched the previous evening's coverage of Gordon Brown's resignation, although BARB figures (published two weeks later) put this at 8.73 million. According to BARB, however, the previous Tuesday's episode of the soap opera *EastEnders* had attracted a slightly higher audience of 9.58 million viewers. Beaten in popularity by mass media entertainment, politics comes to take on the characteristics of that entertainment: indeed the media focus on the subsequent campaign for the election of Gordon Brown's replacement as leader of the Labour Party became what candidate Ed Balls described in August 2010 as a 'daily soap opera.'

Even if the interest in the UK's televised debates had proven a sustainable strategy for a restoration of public involvement in political processes, some have questioned whether the telegenic status of the candidates is necessarily the best way to judge political integrity and ability. Four days before the election Gordon Brown had made a last-ditch attempt to counter the prevailing shift of political culture into the formats of popular television entertainment: 'We're talking about the future of our country. We're not talking about who's going to be the next presenter of a TV game show. We're talking about the future of our economy.' Brown's previous pledges of loyalty to reality television had, however, been recalled by the BBC two days later: 'Gordon Brown is a huge fan of the Simon Cowell brand of reality TV. He has spoken several times about his love of ITV's *X Factor* and *Britain's Got Talent*.' Two days on, Mr Brown lost his parliamentary majority, and five days after that his government was swept from power. It appears the British public were unwilling to heed his belated warning as to the dumbing-down of UK politics.

Realpolitik

In November 2012 it was reported that the Labour Party's Shadow Chancellor Ed Balls's tweets on the subject of *The X Factor* had prompted *X Factor* judge Nicole Scherzinger to ask 'Is he really called Ed Balls?' – before declaring: 'Vote for Balls!' In April 2013 Business Secretary Vince Cable courted greater controversy when he described the astronomical salaries paid to former *X Factor* finalists and boy-band phenomenon One Direction as 'grossly immoral'. In November 2013 Conservative Education Secretary Michael Gove even went so far as to criticize his party's ally Simon Cowell after the latter-day Svengali had suggested that career success was the result of luck rather than of schooling. However in March 2014 Cowell publicly praised Prime Minister David Cameron as a 'decent guy' – albeit one who would make a 'terrible' judge on *Britain's Got Talent*. In the world of politics, reality television has continued, then, to play a curiously significant role.

On 1 January 2010 *The Guardian*'s Marina Hyde had critiqued the Conservative Party's plans to adopt the demagogic strategies of Simon

Cowell by launching plans for an online platform designed to allow the online public to 'resolve difficult policy challenges.' Hyde wrote:

> The Tories have solved the problem of their lack of policies: they are going to wait for the internet to tell them what to do [...] I suspect we are seeing the first instance of Cameron's excruciatingly wrongheaded plan to plunder the oeuvre of Simon Cowell. First TV, now politics – the illusion of deferral to the crowd is the mania of the age [...] Once, there were big ideas in television programming and in government [...] These days, such risks have been jettisoned in favour of allowing the public to think they are writing the script.

Hyde identified a crucial similarity between the persuasive strategies of reality television and those online politics – the illusions of agency afforded to their audiences. Politics is becoming indistinguishable from reality television – and not only in the United Kingdom. In March 2010 the BBC's North America editor Mark Mardell described U. S. politics as increasingly resembling the extremes of reality television. As Barack Obama had commented – during an interview on Jay Leno's talk show in March 2009 – 'Washington is a little bit like *American Idol*, except everyone is like Simon Cowell.' Obama's joke rebounded on him: the following week on the same chat show, Cowell suggested that the U. S. President had been upset when the pop guru had subsequently snubbed a dinner invitation: 'I was invited to have dinner with him last week, but wasn't available.' Cowell made it quite clear who wore the trousers in the relationship between politics and reality TV.

Nick Couldry (2010) has suggested that reality television may increasingly be seen as becoming 'a wholly unacceptable form of social management.' But can popular resistance to the hegemony of reality television conversely represent a reborn desire for socio-political agency? In December 2009 a Facebook-based campaign ensured that the *X Factor* winner's single failed to top the UK's popular music charts that Christmas. This, then, was the extent of the triumph of the protest against globalized capitalism at the end of the week in which the Copenhagen summit on climate change had failed to reach consensus upon a protocol which might adequately address an impending environmental catastrophe. The campaign to secure the top spot in the Christmas charts for Rage against the Machine's 'Killing in the Name' gave its participants a sense of agency sufficient to defer their desire for more significant modes of political empowerment. A year later, this brand of political resistance was reduced to silence in the subsequent campaign to beat the *X Factor* winner's single to the top of the charts with a version of John Cage's *4'33"* (known as 'Cage against the Machine') – in which dozens of musicians kept silent for the prescribed period of time. Nor was there much political capital gained by the April 2013 campaign to bring 'Ding Dong! The Witch Is Dead' to the top of the UK charts in the immediate wake of the death of Margaret Thatcher – a campaign whose triumph of style over substance might ironically have been applauded by those very politicians (the likes of Ronald Reagan, Tony Blair and David Cameron) whose own prioritization of personality above policy might once have proven anathema to such anti-Thatcherite protesters.

If the disempowered populace believe that messing around on Facebook or YouTube represents a real form of empowerment, then they will not attempt to seize actual political power. All the better, then, for the traditional institutions of power. The bread and circuses of contemporary media forms afford more than vicarious pleasures; they offer their publics the impression of real agency, a seductive illusion of empowerment whose dissolution its subjects relentlessly resist.

Yet perhaps there is no possibility of empowerment, only the fact of power; no agency, only the illusion of agency. The illusion of interactivity satisfies and therefore sublimates the desire for real civic participation. We do not seek empowerment by resisting external power because we believe we are already empowered. The illusion of agency fostered by reality television, as by the video game, recruits its subject into its ideological hegemony. These processes of interpellation are sustained by the appearance of interactivity, the transfiguration of ordinary people into figures of extraordinary power (celebrities or avatars), and the impressions of democratic participation and historical significance accorded to this process.

The classic tactics of Disneyfication, McDonaldization and Cocacolonization have now been superseded by processes of Cowelling or X-Factorization. Everything, then – all human history – is transformed into this homogeneous media product, universally palatable although often sickly-sweet. As the *Daily Mail* complained (on 6 August 2012) of the coverage of the London Olympics, 'the media winkles out sob stories about competitors — a family bereavement, a previous sporting setback, an injury — and the audience is set up for their triumph of heartbreak, all scored with suitable music. This is the way we live now: as if we were in some continuous reality TV show.'

You've been framed

The paradox of the bourgeois-proletarian relationship is clear enough to the classical Marxist. What has been less clear is why this exploitative relationship continues. Why has the proletariat not cast off the shackles of its exploitation? Why has the revolution not taken place? Why do the subjects of power submit to – even or collaborate in – the processes of their own victimization? Is it because it is psychologically less agonizing to abdicate one's individuality, and to be assimilated within the dominant ideology, its myths and its discourse – even if that system is designed to destroy you?

As Michel Foucault (1991: 26) points out, 'power is exercised rather than possessed.' Power structurations are self-performing and self-perpetuating; societal systematization is determined not by the conspiracies of sharp-suited men in smoke-filled rooms but by the evolution of institutional, economic and ideological conditions. These structurations are, in Bourdieu's terms (1977, 1986), collectively and objectively orchestrated by institutions rather than by individuals. In these terms, we are no more than the vehicles, vessels or tools of Marx's ideologies or of Richard Dawkins's memes: as Marshall McLuhan (2001: 51) supposed, humanity becomes no more than 'the sex organs of the machine world, as the bee of the plant world, enabling it to fecundate.'

It may be suggested that exploited populations embrace the systems of exploitation because they believe that one day they will become the exploiters themselves – that one day they will win the lottery that is capitalism. Yet this theory does not in itself explain why the proletariat should believe themselves capable or deserving of such a socio-economic transfiguration.

The reason that we each think that we will one day rise up through the system is that we have been programmed to think of ourselves in this way. During the 1990s there was a series of television commercials for Britain's national lottery in which a giant golden hand (the hand of God?) appeared from the heavens and pointed its index finger towards a person who had just bought a lottery ticket: 'It could be you!' We buy lottery tickets because we believe that (against all the odds, both astronomical and divine) there is something special about ourselves which means that we will win. Why then do I think *it could be me*? That, argued Louis Althusser, is a matter of *interpellation*. It is, as L'Oréal would say (in one of advertising's most spectacularly overt moments of interpellation), because you're worth it. For Althusser, mass culture calls to us in precisely this fashion; it announces to us that our lives are meaningful, that we are each of us at the centre of the universe – that each of us has 'got talent' – that each has the X Factor. It is only this belief which allows us to serve the state as sane, submissive, functional citizens. We recognize ourselves in our media heroes (from Lara Croft to Sam Bailey, from Superman to Susan Boyle), and therefore dream of our own significance. Just as the angel Clarence creates for James Stewart a fantasy of individual meaningfulness in Frank Capra's *It's a Wonderful Life*, so the process of interpellation posits the passive individual as an active, centred subject: the individual is thus assimilated within the dominant structures of power.

In 2010 Sweden's TV licensing body *Radiotjänst i Kiruna AB* launched an online viral campaign to promote the payment of its licence fee. This campaign involved a video in which a government official calls a press conference, with the eyes of the world upon her, to announce that a hero has come within our midst: 'This person gives us an alternative to uniformity. We owe this person for making an ordinary day into something special.' The official goes on to ask: 'How can you really trust that what we see on TV and hear on the radio is true? How can we be sure that the weak voices are heard and not scared into silence?' Her answer is that it is of course this heroic individual who allows us to ensure 'that our opinions are really our own.' The face of this hero is shown through the video on placards and billboards and in art galleries: and the face is your own. (This is achieved simply by loading your photo onto the campaign's website.) By paying your television licence fee you have heroically assured the continuation of freedom and democracy.

This video takes Althusser's notion of interpellation literally: in order to achieve its ideological ends – in order to construct its viewer as a responsible citizen who pays her licence fee (as a subject of the mass media) – it demonstrates how one may, in doing so, become the central figure in one's own universe, how one may turn out, as David Copperfield said, to be the hero of one's own life.

Liesbet van Zoonen (2004: 20) has suggested that one of the key reasons for the popularity of reality television is the desire to legitimize one's private life as normative and authentic. Yet, of course, reality television only offers its publics a vicarious and illusory validation of their integrity as individuals. In fact, by making the private public, rather than recognising the universality of the authentic individual, it homogenizes individual specificities and denudes privacy of its defining quality, the authenticity which acts as an exemplar for public conduct. Reality television offers the individual viewer the prospect of a transfiguration into an avatar which projects its own individuality onto the wider (public, civic, social and political) world but in effect this illusion of agency merely veils the transformation of the citizen into the media consumer. A parallel process of transformation appears also to be evident in the pervasive structures and practices of social networking online.

CHAPTER 5

Social Networks

In September 2009 the BBC quoted Facebook Vice-President Mike Schroepfer on the company's mission 'to get as much of the entire world on the social network.' Facebook's ambition is overtly directed towards global domination and by implication the homogenization of the contexts and structures of social interaction. In April 2010 the BBC reported on what it described as 'Facebook's bid to rule the web.' On 22 August 2013 *The Independent* observed that Facebook founder 'Mark Zuckerberg plans world conquest.' Writing in *The Observer* on 1 February 2014 – the company's tenth birthday – John Naughton asked whether the website was 'in danger of swallowing the web.' In April 2014 it was reported that Facebook had accumulated a grand total of 1.28 billion active users.

Facebook's own Facebook page announces that 'Facebook's mission is to give people the power to share and make the world more open and connected.' It is unclear whether this strategy includes the use of a public relations company to plant negative stories about its rival Google, as reported in May 2011. On his own Facebook page the site's founder Mark Zuckerberg lists his personal interests as 'openness, making things that help people connect and share what's important to them, revolutions, information flow, minimalism.' In these terms Facebook represents a revolutionarily minimalist notion of information flow, one in which the flow itself, the process of connection and of sharing, and the condition of openness which affords that possibility, signify more, absolutely more, than the content itself, insofar as what is important to the Facebook user, as to its founder, are these processes themselves. Facebook emphasizes the significance of the act of expression rather than what is expressed. Despite its promises of interactivity, its own processes thereby define its users as the subjects of essentially empty and banal monologues.

Petitioning its user for a status update, the empty field at the top of one's Facebook homepage asks: 'What's on your mind?' Not *what are you doing?* – the life of the user may have become too passive, too solipsistic, to countenance such a call to action or to material interaction: the Facebook user becomes a voice devoid of context, one which expresses itself and only itself, a subjectivity defined by and equal to the processes of Facebook use.

There is a group page on Facebook (established by a member who is 'not a part of the Facebook team') which advises users 'How to permanently delete your Facebook account.' It explains itself thus: 'Ever tried to leave Facebook and found out they only allow you to "deactivate" your account? All your personal data [...] will still be saved indefinitely!' The subject of quitting Facebook is indeed considered so significant that in February 2008 *The New York Times* ran a feature on the company's agreement to make it easier for users to delete their accounts after reports that 'some Facebook users who wished to close their accounts had been unable to do so, even after contacting Facebook's customer service representatives.' In October 2009 that Facebook launched a feature offering friends and family of deceased users of the site the option to 'memorialise' the profiles of their loved ones. Nobody, it seemed, could leave.

In May 2010 Danny Sullivan wrote on the Search Engine Land blogsite that he had recently attempted a Google search on the phrase 'how do I' – only to discover that 'How do I delete my Facebook account' was 'one of the top choices.' Sullivan added that 'if you go back to Google and start typing in "del", you get "delete facebook account" as the top suggestion.' (In 2014 one actually has to type the first four letters before this suggestion appears.) In a subsequent blog entry that same month Sullivan had suggested that, although the company was unwilling to provide him with its cancellation statistics, Facebook users may be cancelling their accounts because of concerns over the company's privacy policies. In January 2013 Sullivan added that Facebook's new search function might raise further privacy concerns in that it allowed for the location of information in ways previously unexpected. On 24 April 2014, *Forbes* magazine reported that the roll-out of the latest version of Facebook's Timeline layout had prompted some users to observe that 'the new design explicitly made finding privacy settings more difficult than ever' – and added that 'Facebook has been known to change privacy settings often, rendering these settings confusing, which can affect the way users' data is accessible by other users and advertisers.' A fortnight earlier *The Washington Post* had noted that 'if Facebook and privacy had a relationship status, it might be *It's complicated*.' The same month an online marketing campaign for the video game *Watch Dogs* demonstrated that it could draw upon publicly available Facebook information about a user to construct a dossier on that individual's location, contacts, interactions, online activity patterns and other personal details – all reassuringly compiled in the style of an assassin's target profile. (Cf. Vincent 2014b.)

The following month Facebook announced plans to serve advertising on third-party mobile apps. The BBC reported that, despite privacy concerns, the company enjoyed significant commercial advantages in this area thanks to 'the depth of knowledge the firm has about its users.' Facebook's vision of openness and connectedness might be interpreted in terms of a data-mining strategy which commodifies its users as packages of commercially valuable information. Oymen Gur (2010) has suggested that 'even though social networking sites seem transparent, users are still mediated through them – the more people are liberated with wider and more transparent networks, the more they are constrained.'

Boyd and Heer (2006) have described online social networking as a process through which individuals write themselves into being. Siibak (2009) argues that selfhood is constructed in social media sites according to ideals or expectations of selfhood. Sites like Facebook offer their users a mode by which to 'capture and share [their] life story' (Kelsey 2010: 1). Updates to the site's structure led the BBC to suppose in September 2011 that 'identities will now be defined through a densely packed vertical timeline of major life events.' The BBC quoted technology research analyst Sean Corcoran as supposing that 'Facebook is positioning itself as your life online.'

Eager to report the most enthralling status updates, it is not perhaps uncommon for Facebook users, like diarists or autobiographers, to lead their lives and frame their identities according to the needs of their narratives. Paul de Man (1984: 69) wrote that 'we assume that life *produces* the autobiography as an act produces its consequences, but can we not suggest, with equal justice, that the autobiographical project may itself produce and determine the life and that whatever the writer *does* is in fact governed by the technical demands of self-portraiture and thus determined, in all its aspects, by the resources of his medium?' Roland Barthes (1977: 144) suggested of Marcel Proust that 'he made of his very life a work for which his own book was the model' and the users of social media may be performing similar practices. Although YouTube may have adopted the slogan *Broadcast Yourself*, it is clear that many of that site's users are intent upon inventing new selves for themselves – or even, as in the case of such YouTube stars as EmoKid21Ohio or Lonelygirl15 (both videobloggers famously exposed as fictitious characters), upon counterfeiting such selves. These extreme attempts to forge identities offer fitting metaphors for the processes of self-invention that to some extent underpin all acts of autobiography.

In February 2013 it was reported that a 104-year-old woman had been obliged to lie on Facebook about her age: when she tried to input her true date of birth she found that the site refused to let her 'enter in a date that goes back that far.' The following month (on 12 March 2013) *The Daily Telegraph* reported the findings of a OnePoll survey which suggested that a quarter of women regularly fabricated activities in their social media updates: 'they mostly pretended to be out on the town, when in fact they are home alone, and embellished about an exotic holiday or their job.' The next day the paper's Emma Barnett pointed out that the survey in question had only been targeted at women, and that therefore we might also surmise that men lie on social media too: 'why would anyone want to project a negative image of themselves on networks designed to make people brag about their lives?'

The status update thus threatens to become the end – both the goal and the ruin – of one's social life. Samuel Johnson's advice to the prospective diarist (Johnson 1951: 176) remains timelessly appropriate when he cautions against 'a great part of life be[ing] spent in writing the history of the rest.' One could do worse than to heed the wisdom of Richard Ayoade's Maurice Moss in the 'Friendface' episode of Graham Linehan's *The IT Crowd* (2008) as he explains his resistance to participation in social networking sites: 'I think I've got better things to do than talk to friends and flirt with people, thank you very much.'

Greenfield (2013) has argued that the Facebook self – defined by 'the instant thumbs-up from others' – may eventually come to 'mean living a life where the thrill of reporting [...] completely trumps the ongoing experience itself.' She has predicted that 'if we're going to be living in a world where face-to-face interaction, unpractised as it is, becomes uncomfortable, then such an aversion to real life [...] may be changing the very nature of personal relationships themselves.' The nature of the interpersonal relationships which comprise society of course determines the very nature of that society itself.

Status symbols

Why might Facebook users so relentlessly endeavour to lead lives worthy of their status updates? Why might they interrupt the flow of their real-world interactions in order to post such updates onto the site? Although this activity may in part relate to an internal process of self-narrativization – a coming to subjective integrity through the reconstruction of one's life within the logic and aesthetic of the narrative form – it also represents an expression of one's own public identity and social status.

While the status update announces and performs the identity of the user, it is specifically the user's number of Facebook friends which provides material evidence for the success and status of that user's constructed identity. Danah Boyd (2006: 13) supposes that one's collection of Facebook friends affords a space for such identity performance.

Boyd (2006: 10) proposes that 'by having a loose definition of Friendship, it is easy to end up having hundreds of Friends [...] Because of how these sites function, there is no distinction between siblings, lovers, schoolmates, and strangers. They are all lumped under one category: Friends.' One might also advance the argument that it is only possible to accumulate these hundreds of friends by changing one's definition of friendship. The social networking site seems, as Papacharissi (2010: 55, 63) suggests, a social realm which, while more fluid than traditional modes of socialization, appears fragmented, superficial and commodifying.

On 2 March 2014 Danah Boyd told *The Daily Telegraph* that 'the heavily scheduled nature of their lives, plus their lack of physical mobility' have made 'face-to-face interactions increasingly impossible' for many young people. She proposed that those young people who appear to have developed inseparable symbiotic relationships with their mobile telephones 'aren't compelled by gadgetry, they're compelled by friendship; the gadgets are primarily a means to a social end.' It is unclear why Ms Boyd believes that contemporary youth have suddenly become less mobile than previous generations – to the extent that in-person social interactions have become impossible. She may be right, however, when she suggests that it is not the gadget itself which is important: what matters is the easy mediation which that gadget affords – the detachment of interpersonal relationships from immediate and material commitment.

Matthew Tedesco (2010: 123) has argued that 'our ordinary understanding of friendship is of a relationship, where that relationship requires something of us. We spend time with our friends; we invest ourselves in our friends. But interactions with our Facebook friends usually aren't like this […] our exchanges are often ephemeral and impersonal, lacking any of the investment that we find in our ordinary friendships outside of cyberspace.' To what extent then can we call this strange, new, virtual thing by that old name of friendship?

The Geography of friendship

Samuel Johnson's definition of the word *intimate* in his *Dictionary* of 1755 proposes, amongst other things, that the intimate is 'near; not kept at a distance.' Robin Dunbar (2010) has argued that, although social networking sites may have value in supporting ongoing offline relationships, relationships conducted entirely via such sites are neither as stable nor as strong as relationships which take place in the material world 'because you're not interacting face to face.' There therefore appears to be a real correlation between physical and emotional closeness. As Aristotle (2004: 210), one of western civilization's first theorists upon the nature of friendship, argued, friends who do not spend much of their time together are not really friends.

Meditating on Aristotle's position, Jacques Derrida (2005: 222) has argued that 'if absence and remoteness do not destroy friendship, they attenuate or exhaust it.' Derrida goes on to point out that Aristotle's argument did not allow for the possibility of friendship mediated by telecommunications. It was not that Aristotle ignored the possibility that friendships might be sustained through contact via media of telecommunications because such media did not exist in his time (they did: there were, for example, such things as letters); Derrida implies that he ignored these possibilities because he did not consider them a viable way to maintain friendships.

New technologies have not abolished distance, although they can allow us to forget its enduring significance. When we interrogate the notion of online relationships, geography seems both a banal and an essential factor in the ongoing shift in the concept of friendship. The moral aspects of traditional notions of social and emotional intimacy seem dependent upon the material conditions of that intimacy. From this perspective the limits to the possibilities of virtual friendship appear insurmountable. Craig Condella (2010: 114) has asked whether 'the friendship spoken of by Aristotle is altogether different from the virtual friendships formed on sites like Facebook.' He returns in this context to the Aristotelian need for physical presence, and this, he argues, is where Facebook cannot deliver.

It's complicated

Boyd (2008b: 211–212) writes that friends on social networking sites are not necessarily equivalent to friends in physical life: 'participants use the term *Friends* to label all connections, regardless of intensity or type.' Boyd (2008a: 17) proposes that social networking sites may foster a 'fake sense of intimacy' with people one does not really know. Boyd (2006: 10) therefore suggests that, within the social networking site, 'there seems to be little incentive to be selective about Friendship [...] As far as most participants are concerned, Friendship doesn't mean anything really.' One might therefore ask whether the non-selective view of online friendship (if friendship doesn't mean anything) is changing traditional views of friendship in offline interaction – or whether users are able consciously to distinguish between these modes of online and offline relationships?

Danah Boyd has said that 'fundamentally people know the difference between a Friend (as in social network sites) and a friend (as in the intimate connection). This is why I go out of my way to differentiate them.' The obvious question is why one would need to go out of one's way to differentiate these modes of friendship if these distinctions were truly so clear. The globalizing structures of such social networking sites as Facebook are imposing a homogeneous notion of friendship upon the world; and insofar as users increasingly interact in a primarily online environment, their ability to distinguish between online and offline modes of friendship seems less and less relevant. As the phrase 'social networking' has come almost exclusively to refer to *online* social networking, is it so difficult to imagine a time when the word 'friend' might refer primarily to online friendship? Indeed, when Boyd (and others) view the Facebook friend as a 'Friend' (with a capitalized 'F') in an attempt to differentiate this phenomenon from offline manifestations of friendship, this translation of the term into a proper noun may suggest there is something absolute, definitive or seminal about this mode of friendship (in parallel, say, to the differentiation of 'God' from 'god' – or of 'Internet' from 'internet').

According to Facebook's own statistics, the average user spends about an hour on the site every day. This suggests that, even if Facebook users remain able to distinguish between modes of online and offline friendship, many are committing significant social investment to their online relationships. Indeed, a study reported in *The Daily Telegraph* in November 2010 announced that one in four Britons socialises more online than in person. Smallwood (2010) has meanwhile reported the extreme example of one Facebook-addicted family in which children no longer made eye contact or talked with each other. One is tempted to imagine that if the internet were ever irreparably broken (admittedly an unlikely scenario) there might be many young people who no longer had the social skills to go out and meet people in physical environments: one can envisage, in this absurd scenario, youngsters sitting in bars handing each other notes revealing their status updates and requesting friendship.

A member of Facebook's Data Team, Adam Kramer, has however commented:

> It's important to remember that the vast majority of friendships on Facebook are actually traditional relationships that happen to also have a Facebook representation which is termed 'friend'. As far as friendship-like social dynamics are concerned, there's a decent amount of empirical evidence suggesting that communication through Facebook serves to strengthen weak-tie bonds, or relationships among people who are not very close. But most broadly, it is my belief that Facebook and social networking sites are a facilitator of social interactions, and do not indicate anything new about the psychology of human interaction. Rather, they serve as tools to encourage social interaction in new ways via new means of communication, much like the printing press, telephone and television did in their day.

Yet, without wishing to sound overtly technologically determinist, one might suggest that Gutenberg, Bell and Baird's inventions did more than facilitate and accelerate communication; one need not be a disciple of Marshall McLuhan to imagine that these media technologies radically affected the nature and content of social interactions.

When new media technologies (such as the internet and the mobile telephone) converge, one would expect their impact to be all the more intense. In 2013 approximately 945 million users (more than three quarters of all active users) accessed Facebook through mobile devices. The engineer who designed the hugely successful 'Facebook for iPhone' package, Joe Hewitt has commented that 'Facebook hasn't affected contemporary notions of friendship significantly. Unlike some other social networks where one is encouraged to "friend" people they've never met, Facebook works best when it is a mirror of the real world, and your friends are people you know in real life.'

Both Kramer and Hewitt see Facebook as merely a facilitator for traditional offline friendships. The medium is, it seems, no longer the message; the nature, content and meaning of a communication act are no longer affected by that act's own conditions of process and possibility. Burke, Marlow and Lento (2010) have similarly supposed that social networking sites merely complement real world relationships. However, Moira Burke has more recently added:

> Social networking sites certainly changed the threshold and nuance that most people experienced in the face-to-face definition of 'friend'. This is primarily because they forced people to make relationships explicit, and 'friend' status was often simply an access control mechanism for content. All kinds of social roles were collapsed into a single term. Between the awkwardness of having to explicitly deny someone's friend request and the publication of one's friend count on the profile, many users hoarded friends, in a socio-economic way.

In Judd Apatow's 2009 film *Funny People* Adam Sandler commented that the more friends you have online, the fewer friends you have in real life. One might at least suggest that one's focus upon the modes of friendship promoted and defined by social networking sites may change the nature of one's relationships offline, and that the amount of time spent involved in these online processes of social interaction, the number of 'friends' one has online and the emphasis one places not only upon these modes of interaction but also upon the significance of the quantities of friendships achieved and maintained online may radically affect one's notion of friendship itself.

Robin Hunicke (2008) has suggested that Facebook allows its users to feel that they are living enjoyable lives and that they are loved: 'Facebook makes people feel like they matter.' As such, one might argue that it offers not the liberating pursuit of happiness but the sentimental impression that one might already be happy and that this is perhaps the extent of what happiness is. This is, of course, yet another illusion of empowerment which pre-empts and prevents the struggle for actual social, political and cultural power.

The commodification of friendship

An exercise has been conducted to test the extent to which Facebook users exercise discrimination in their acceptance of friends. A dummy Facebook account was created under an Anglophone name which, although ordinary-sounding, did not exist on Facebook – nor (according to a Google search) anywhere on the internet. This name was chosen specifically to avoid the possibility that it might be mistaken for that of a real-life friend. The Facebook account did not include any images of the account-holder nor any profile information. In stage one of the exercise 100 Facebook users were sent friend requests: the names of these users were generated randomly. Within five days 18 per cent of these Facebook users had accepted the dummy user as a friend.

In the second stage of this exercise a further 100 friend requests were sent to Facebook users. These recipients were all contacts generated by Facebook itself as suggested friends – in other words they were friends of those who had already accepted the dummy user as a friend. Within five days 74 of these friends of friends had befriended the dummy user.

In the third stage of this exercise a further 100 friend requests were sent to Facebook friends of those who had accepted Facebook friendship in stage two (i.e. friends of friends of friends). Within five days 84 of these friends of friends of friends had befriended the dummy user.

These results appear to demonstrate that while the majority of Facebook users are relatively discriminating in accepting the friendship of unknown contacts, the majority are also willing to accept friendship if they know someone who knows the person requesting friendship. Indeed, they appear to be even more likely to accept that friendship if more than one friend is already shared, and if the petitioner already possesses a significant number of friends. It appears that any friend of anyone who is already a friend of mine is also a friend of mine. This mode of friendship is ostensibly one based entirely upon quantifiable notions of social capital: the quantity of social capital possessed by the seeker of friendship (the number of mutual friends or their total number of friends) is directly related to the probability of the success of their petition. The petitioner's perceived level of social capital may therefore be seen as adding to the social capital accumulated by the petitioned party. This inference is further evidenced by the fact that during the second and third stages of this exercise the dummy account received seven unsolicited requests for friendship from friends of friends who had not been contacted – as this non-existent user accumulated social capital he came to be perceived by third parties as a commodity worth investing in. In Aristotle's terms this is therefore a mode of friendship based upon utility rather than upon virtue or moral commonality.

Burke et al. (2010) have argued that the uses of social networking sites (and specifically friend counts) have been associated with increased levels of social capital. Ellison et al. (2007: 1161) have also noted a correlation between Facebook use and the accumulation of social capital (cf. Zywica and Danowski 2008; Valenzuela et al. 2009) – while Papacharissi (2010: 139) has also noted that online communication has been associated with the creation of social capital. More recently, the likes of Sajuria (2014) have continued to argue that the formation and distribution of social capital are central to the function of such social networking sites as Facebook.

Fenton (Curran, Fenton and Freedman 2012: 142) has written of the arithmetical commodification of the economic value of Facebook friendship: 'the numbers of friends you have on Facebook [...] are markers of success.' Freedman (Curran, Fenton and Freedman 2012: 82) has added that friendship 'becomes the currency that drives the network. He has argued, however, that 'for many people who have no wish for their friendships to be privatised on Facebook or for their personal data to be surveilled and sold on Google, this is a form of commodification in which their very labour, their own creative self-activity, is repackaged and turned into an object to be exchanged, at a price, on the open market' (2012: 83).

The utilitarian commodification of friendship was depicted by Facebook's founder Mark Zuckerberg as the website's primary function when he spoke in April 2010 of Facebook's capacity to allow its users to 'use' their friends: 'If you look back a few years ago and even as recently as today, in most cases the web isn't designed to use your friends. We want to be one of the things that empowers that and right now most users are using Facebook.'

If friends are a commodity whose quantity enhances one's level of happiness, they are clearly one to be valued, and one whose integrity is to be regulated by the appropriate corporate authorities. In November 2009 it was reported that Facebook was threatening legal action against a company called USocial that offered to sell its users' friends and followers on Facebook and Twitter. (On Twitter, for example, 1,000 followers cost £53.) In May 2012 it was reported that Facebook was piloting a system whereby users could pay a fee to increase the prominence of their posts. Friendship has clearly become an economic commodity whose quantification is a crucial concern. In a Facebook blog entry of April 2009 Facebook's Chief Operating Officer, Sheryl Sandberg wrote that 'when our Data Team measured active networks for users on Facebook, it found that, in any given month, users keep up with between 2 times and 4 times more people than through more traditional communication.'

A technology which can increase one's quantities of friendship – and therefore one's levels of happiness – fourfold is clearly one which is to be cherished. A number of key questions however arise. When one increases one's number of friendships fourfold is each of those friendships as valuable as when one had fewer friends? Can friendship exist in a virtual environment which excludes the degree of material (spatial and temporal) commitment and even sacrifice which in the offline world tests, tempers and defines it? When friendship becomes a quantifiable commodity and a reserve of economic capital, is it still friendship? When we count our friends, do they count as friends? And, if not, might this increasingly bankrupt concept of online friendship return to undermine the integrity of friendship in the offline world?

Such news stories as 'Facebook gatecrashers wreck family home at teen party' (*Daily Telegraph*, November 2008), 'Family home trashed after Facebook party goes wrong' (*Daily Telegraph*, February 2010), 'Facebook party invite sparks riot' (*BBC News*, 22 September 2012) and '£30,000 trail of destruction [...] after girl, 14, advertises her party on Facebook' (*Daily Mail*, 10 December 2012), or the tale of 'the teenagers who organised a Facebook party that descended into a street brawl' (*Daily Telegraph*, July 2009), or the account of 'the German teenager who advertised her 16th birthday party on Facebook only for 1,600 gatecrashers to turn up at her house' (*Daily Telegraph*, June 2011), do little to promote a perception of the ideals of Facebook friendship. In August 2009, Vincent Nichols, head of the Roman Catholic Church in England and Wales, expressed his concern that sites such as Facebook 'encourage teenagers to view friendship as a commodity.' The Archbishop added: 'It's an all-or-nothing syndrome that you have to have in an attempt to shore up an identity; a collection of friends about whom you can talk and even boast. But friendship is not a commodity, friendship is something that is hard work and enduring when it's right.'

A similar phenomenon whereby one's number of friends or 'followers' determines one's social and cultural status may also of course be witnessed on Twitter. As Jerome Taylor wrote in *The Independent* newspaper in August 2010: 'You might be the 21st century's Oscar Wilde, with an acerbic wit of such magnitude that you can distil the world into razor-sharp aphorisms of just 140 characters. But unless you have people following you, no one will even know you exist.'

The quantification of friendship reimagines the ideal of friendship as a saleable commodity. This commodification negates the very concept of friendship itself, one from which such seminal thinkers as Aristotle (2004: 204), Cicero (2010: 14) and Montaigne (1991: 207) specifically exclude those so-called friendships based upon individual utility or benefit. Utilitarian friendship represents a lesser, shallow or empty version of the ideal.

The Arithmetic of friendship

Dunbar (2010) has suggested that the accumulation of Facebook friends has become competitive, and argues that those with hundreds of such friends probably do not know most of them. Aristotle similarly suggested that the number of friends one can maintain is limited by the number of people with whom one can sustain intimacy, and therefore proposed that those who boast many friends are in fact 'friends of nobody.' (Aristotle 2004: 251). In his reading of Aristotle's notions of friendship, Jacques Derrida has argued that when you try to quantify friendship as an economic commodity (and therefore when you try to increase your number of friends in order to increase your social capital, the utility adhering to those friendships) then they are no longer friends in Aristotle's sense. Derrida (2005: 21) suggests that friendships are not objects or numbers, and that friendship itself resists quantification.

Derrida's analysis of the history of the concept of friendship in western civilization turns to the work of the sixteenth century French essayist Michel de Montaigne to argue that this concept requires both intimacy and exclusivity. Montaigne (1991: 215) speaks of the purity of that friendship in which 'each gives himself so entirely to his friend that he has nothing left to share with another.' One of the most celebrated of friends in the history of letters, the lexicographer Samuel Johnson's long-suffering companion James Boswell (1986: 247) also saw friendship as discriminating and exclusive: he supposed that one must prefer the interests of a friend against those of others and endorsed the Aristotelian notion that 'he that has *friends* has no *friend*.'

Can such intimacy, requiring such exclusivity, be possible in the resolutely public space of Facebook? Facebook's attempts to balance the contradictory imperatives of intimacy and openness (to some extent determined by the needs of its users and of its advertisers respectively) may not be sustainable. In May 2012 the BBC's Rory Cellan-Jones noted that Facebook founder Mark Zuckerberg had not listed his wedding that month as an upcoming event on his Facebook page and had therefore not been bombarded by Facebook adverts from cakemakers, florists and bridal couturiers. Cellan-Jones went on to ask whether therefore Facebook's model of 'frictionless sharing' is sustainable when even its own founder seems unwilling to share his personal details. As David Kirkpatrick (2011: 333) has suggested, 'reciprocal personal connections packed with very private data may not coexist well with unbridled sharing.' Kirkpatrick goes on to ask whether a system so deeply founded upon personal trust could ever really be transparent and open. The ideology of Facebook – its mission of uncompromising and obligatory openness, its removal of barriers and of distinctions – makes this seem unlikely. When everyone is special, no one is.

Boyd (2006: 3) argues that 'the category of friend carries an aura of exclusivity and intimacy.' Aristotle, Montaigne, Boswell and Derrida would agree. A corollary of this proposal is evident: that the Facebook contact is, within such traditional definitions, not a friend as such. The meaning of friendship is therefore in the process of changing: it is not only that the *concept* of Facebook friendship (a new and radically different sense of friendship) is overwhelming that of offline friendship, it also seems that the *practice* of offline friendship is being overtaken by the practice of friendship online.

The ideal of friendship has seemed essential, for several significant European thinkers, to the perpetuation of democracy itself. Aristotle (2004: 220), for example, supposed that 'while in tyrannies friendships [...] are little found, they are most commonly found in democracies because the citizens, being equal, have much in common.' Friendship is as such a prerequisite for democracy: as Derrida (2005: 22) has pointed out, for Aristotle at least 'there is no democracy without the community of friends.'

Papacharissi (2010) suggests that the maintenance of the private sphere sustains the possibility of a broader socio-political public sphere (in that the public is founded upon the private) – but that, insofar as social media blur the distinctions between the private and the public, the dissolution of the integrity of the private sphere also undermines the integrity of public interactions. Derrida has argued that the concept of friendship lies at the core of the western political tradition, and specifically at the heart of the notion of democracy: democratic trust and interactivity is extrapolated from the ideal of friendship; and therefore Plato's ideal of democracy grows out of Aristotle's ideal of friendship. The absolute and irreducible discrimination of friendship stands as a benchmark by which to measure our broader social and political relationships.

If all the world were Facebook friends, would that then spell the end of 'real' friendship, and even of democratic politics? The resulting 'world without politics' – as Derrida (2005: 130) put it – would appear to have been 'abandoned by its friends' and seem little more than 'a dehumanized desert.' We return yet again, then, to Baudrillard's desert of the real – the waste land of the new, virtualized reality.

The revelation of secrecy

Among its various glosses thereof, Samuel Johnson's *Dictionary* (1755) defined an *intimate* as 'one who is trusted with our thoughts.' Michel de Montaigne (1991: 215) argued that the intimacy of friendship could permit no indiscretion: 'that secret which I have sworn to reveal to no other, I can reveal without perjury to him who is not another: he *is* me.' Yet when we publicly tag our Facebook friends in compromising photographs – when we simultaneously announce and denounce our intimacy – we seem further from Montaigne's idealism than from Friedrich Nietzsche's typical cynicism on the subject: 'There will be but few people who, when at a loss for topics of conversation, will not reveal the more secret affairs of their friends' (Nietzsche 2008: 171).

Derrida (2005: 62) points out that, for Nietzsche, friendship displays the semblance of intimacy but lacks 'actual and genuine intimacy.' Friendship is therefore a matter for Nietzsche not of intimacy but of intimation and innuendo. Aristotle and Montaigne create a realm of silence around friendship (nothing is intimated beyond the limits of its intimacy) which allows an uncompromised and unassailable mode of communication within the friendship; Nietzsche, by contrast, permits external intimation (which is the betrayal of intimacy), but cannot countenance such openness *within* the intimacy.

Nietzschean friendship, like Facebook friendship, is therefore a matter of utility which depends upon the intimation of a relationship and upon the simulation of selfhood. 'Is it in honour of your friend that you show yourself to him as you are? [asks Nietzsche (1969: 83)] He wishes you to the Devil for it!' Friendship is for Nietzsche sustained not by openness and intimacy *per se* but by maintaining the pretence of these things. This mode of friendship is not, however, sustainable in itself, because it is unequal, self-serving, arbitrary and indiscriminate. It recalls an intimacy censured by Samuel Johnson (1971: 584): 'his friendship was [...] of little value [...] because he [...] would betray those secrets which, in the warmth of confidence, had been imparted to him.'

This Nietzschean notion of friendship as a public and treacherous intimation of trust may not be entirely unfamiliar to the user of Facebook (to those for example who tag their friends in embarrassing photographs) but seems antithetical to the ideal prescribed by Aristotle, Cicero, Montaigne and Derrida, a mode of friendship described by Maurice Blanchot (1997: 291) as one 'which does not allow us to speak of our friends, but only speak to them.'

The loss of virtue

In August 2011 the *Liverpool Echo* reported that the Facebook page of the murderer of teenager Anthony Walker contained 'the same racist language that was used to taunt Anthony in the moments before his death.' In September 2011 *The Guardian* reported that Facebook user Sean Duffy had been gaoled after posting onto the site messages mocking the deaths of teenagers and taunting their families.

In July 2007, during a family holiday in the Algarve, Madeleine McCann went missing just a few days before her fourth birthday. On Christmas Eve 2009 the *Daily Record* reported that:

> A sick Facebook group claiming to be created by Maddie McCann's kidnapper has been removed after thousands complained. The group named 'If 2,000,000 people join this group, I will give back Maddie McCann' was slammed by outraged users who campaigned to get it banned.

In March 2009 35-year-old Claudia Lawrence also went missing. On 18 February 2010 *The Northern Echo* reported:

> A sick hoaxer who raised false hopes in the long-running investigation into the disappearance of Claudia Lawrence has been arrested. The teenager admitted posting a bogus message on the social networking site Facebook [...] It read: 'Hi everyone just let you be aware that I am ok and I am safe and sound. Speak to you all soon. Claudia xxx'

We may note in both reports the use of the word *sick*: Through such much-reported cases Facebook comes to seem the realm not of real friends but of morbid and malicious spectres, impostors and doppelgängers, a virtual parody of the relationships it purports to reflect and sustain. Aristotle (2004: 200) wrote that 'friendship is a kind of virtue, or implies virtue.' What then when that virtue is lost?

In October 2012 a man from Lancashire was charged with making an offensive post on Facebook about a missing five-year-old child. In March 2013 it was reported that a woman had received a suspended prison sentence after claiming on Facebook to be someone who had gone missing as a toddler in 1981. When in June 2012 a woman who had lost her baby during pregnancy started a campaign to close a Facebook group which shares jokes about dead babies, Facebook was reported as refusing to do so on the grounds that 'it did not break any rules.'

In May 2013 Facebook announced that it would review its policies on 'controversial, harmful and hateful content.' That same month the company rescinded its previous refusal to delete videos of decapitations; but in October 2013 it reversed this policy again, prompting British Prime Minister David Cameron to respond (inevitably through Twitter) that he considered it was 'irresponsible of Facebook to post beheading videos, especially without a warning.' Tweets in response to Mr Cameron noted (1) that Mr Cameron may have miusunderstood the situation – i.e. that it was not Facebook itself which was posting such videos, and (2) it might be better to delete such videos rather than merely to include a helpful warning. (Mr Cameron or his electronic amanuensis had failed to regale his tweetees with his wisdom as to how such a warning might be worded.) In January 2014 many were again outraged when an image of the murder of a British oil executive on a Libyan beach appeared on Facebook.

In September 2012 a British teenager was found guilty of posting an offensive Facebook message about the deaths of six British soldiers in Afghanistan. In February 2013 an investigation was launched into allegations that a paramedic had written on Facebook that he hoped a particular hospital campaigner would suffer a 'life-threatening illness.' In March 2013 it was reported that a video showing an act of apparent child abuse had gone viral on Facebook. A Belfast councillor was charged in August 2013 after making a 'grossly offensive communication' on Facebook about Irish Republicans. In April 2014 a 17-year-old girl from South Wales was arrested on suspicion of posting 'grossly offensive' comments on Facebook in relation to the death of a local 17-year-old boy. The local police released a statement reminding the public of their 'responsibilities to keep comments on social media within the law.'

When, in August 2010, a middle-aged British woman called Mary Bale was caught on CCTV dumping a cat into a wheelie bin, the social media outcry seemed rather less focused on feline welfare than upon acts of revenge. Facebook erupted with such pages as 'Mary Bale should be locked up for

putting Lola the cat in a bin', 'Let's all lock Mary Bale in a bin and leave it in
a deserted place', 'Mary Bale Named and Shamed', 'Cats unite against Mary
Bale', the 'Mary Bale Hate Group', 'Mary Bale – Stinky and Stale', 'Damn
Mary Bale', 'Mary Bale is a cruel bitch', 'Mary Bale – Heartless Old Hag',
'Mary Bale is an Idiot Psycho Bitch', 'Mary Bale is EVIL', 'Death penalty
for Mary Bale', 'Death to Mary Bale' and at least twenty other variants on
this theme. The 'Death to Mary Bale' page was removed after one Facebook
user had declared that the woman should be 'repeatedly head butted' whilst
another wrote that 'she should be flogged to within an inch of her life.'

While these instances remained at a level of verbal abuse, uses of the
social networking site have also led, both directly and indirectly, to more
direct forms of violence. It was, for example, reported in April 2010 that
Londoner Awais Akram had been stabbed and had acid thrown in his face
in revenge for an affair he had conducted with a married woman he had
met on Facebook.

Facebook advances a public exposure of intimacy which might have
devastating consequences upon the relationships involved. According to
research published by Divorce Online in May 2012, Facebook use had been
implicated in a third of all divorce proceedings. A study published by the
University of Missouri is June 2013 also found that Facebook use may cause
'negative relationship outcomes including emotional and physical cheating, breakup and divorce.' In a 2010 episode of Graham Linehan's situation comedy series *The IT Crowd*, one character, unwilling to tell his wife
directly of his plans to divorce her, instead changes his Friendface profile
from 'married' to 'it's complicated' in the hope that 'someone will pass it on.'
Earlier, in a 2008 episode of the same series, a husband's observations of his
wife's social networking site activities had led him to suspect her infidelity:

> I think Deline might be sleeping with someone else. Little clues here and there. Things
> only a husband would pick up. Like someone wrote a message on her Friendface
> wall. Read: 'Can't wait to shag your arse off again soon.' I don't know. Maybe I'm
> reading too much into it.

Others have responded to such revelations rather more harshly than
the husband in Linehan's comedy. In November 2008 Wayne Forrester
was gaoled for stabbing his wife to death – as a result of her changing her

relationship status on Facebook to single. In September 2009 Welshman Brian Lewis was gaoled for a remarkably similar crime: stabbing his partner to death after she had changed her Facebook relationship status to single. Two months later John McFarlane was convicted of murdering a woman with a bolt gun shortly after she had posted a message on Facebook describing his belief that they might have a relationship as delusional. In March 2010 Paul Bristol was found guilty of murdering his former girlfriend after seeing a photograph of her and her new boyfriend posted on Facebook.

The case of Raoul Moat, a former panel-beater and tree surgeon from Newcastle upon Tyne who sparked a massive police manhunt in July 2010, was also conspicuous for its Facebook connections. The news media noted that Moat's shooting spree appeared in part to have been catalyzed by his use of Facebook, and was anticipated by his own Facebook status update: the grandmother of his former girlfriend (and one of his victims) told journalists that Moat had threatened his ex with a gun 'all because she'd put on her Facebook that she was going out with a friend.' The press also noted that immediately before he launched his attacks Moat had updated his Facebook status to read: 'Just got out of jail, I've lost everything, my business, my property and to top it all off my lass has gone off with someone else. Watch and see what happens.' Following Moat's death at his own hands at the climax to a protracted police manhunt, Prime Minister David Cameron condemned public shows of sympathy for the deceased gunman left on Facebook. (Cf. Williams 2014.)

Facebook often seems to have become an arena for the perpetuation of acts of extreme violence. In January 2009 a teenager called Leon Ramsden was convicted of murder, having stabbed a man to death just hours after posting onto Facebook that he felt like killing someone. In April 2012 a man who had claimed he was going to commit a 'live murder on Facebook' was gaoled for two years. December 2013 saw a Miami man charged with murdering his wife and then posting a photo of her dead body on Facebook.

In his anthropological study of the impacts of Facebook upon the Caribbean island of Trinidad, Daniel Miller (2011: 11) notes that people 'are always alert to [...] any change in the relationship status that is posted so conspicuously on the profile of every Facebook account.' Among his case studies, Miller tells the story of a Trinidadian who blames the collapse of his marriage on the pervasiveness of Facebook. Miller (2011: 12) writes that 'Facebook has turned everything into confusion, into public slavering and gossip [...] that sours the relationship itself. It has lost its protection of intimacy and shared secrets.' The virtualization and publicization of relationships on social networking sites appears to undermine, confuse or corrupt the intimacy of such interactions. The case of Victoria Jones, the teacher who was reported in September 2011 as having stolen baby pictures from Facebook in an attempt to make her ex-boyfriend think they had a baby, is a particularly extreme example of this phenomenon. As, for that matter, is the case of another British teacher who was, according to news reports of February 2012, reprimanded by her employers after comments about her social life were seen by pupils on her Facebook page.

Facebook effaces the possibility of privacy. Facebook forces all of one's hubris, anger, humiliation and loss into the glare of public scrutiny: it performs, as Papacharissi (2010: 25–50) suggests, a process characteristic of the workings of new media technologies whereby the traditional boundaries between public and private are blurred. The experience of Charlotte Fielder (as reported in June 2011), an amputee who discovered that her Facebook profile picture had been posted onto a pornographic website, is a clear, albeit extreme, case in point – as indeed is the story of Sir John Sawers who, shortly after being appointed head of Britain's secret intelligence service MI6, was embarrassed in July 2009 to find that his wife had posted family details and photographs (with limited privacy settings) onto her Facebook account. Two years later, in June 2011, the British Ministry of Defence launched a campaign to discourage its personnel from sharing sensitive information on Facebook.

In December 2009 police were reported as using Facebook status updates in their attempts to track down convicted burglar Craig Lynch, an escaped prisoner who was regularly informing his Facebook friends about the events of his life on the run: from a near car-crash on an icy road to the details of his meals. The following February another burglar, Roy Boodle, was gaoled for three and a half years after having used his mobile phone to taunt detectives with Facebook messages during his 18 months on the run from the Greater Manchester Police. It sometimes feels like these miscreants simply cannot help themselves: as if the addictive pleasure of online self-promotion outweighs all other practical considerations. In April 2014 a Cheshire burglar was arrested after posting onto Facebook a picture of himself posing with an air rifle he had stolen. The owner of the gun had recognized his property on the social networking site.

Facebook activity not only provokes, motivates or reveals enmity; it may in fact become a tool in the planning and execution of crimes of intimidation and violence. In February 2009 George Appleton was found hanged following the discovery of the badly burnt body of his former girlfriend; police noted that Appleton had been known to contact women through such social media sites as Facebook. In November that year Keeley Houghton had become the first person in Britain to receive a custodial sentence for online bullying – spending six weeks in a young offenders' facility after posting a threatening message about another young woman on Facebook, a message which announced that she was 'going to murder the bitch.' The following February it was revealed that organized crime boss Colin Gunn had, while in prison, declared on Facebook: 'I will be home one day and I can't wait to look into certain people's eyes and see the fear of me being there.' A week later Justice Secretary Jack Straw had announced that 30 Facebook pages had been removed because they had been used by convicted prisoners to taunt their victims.

In March 2014 the *Telegraph* reported that a violent criminal had 'taunted his victims by posting a picture of himself topless on a Thai beach.' The *Express* reported in March 2013 of a convicted rapist's attempt to frame his victims via Facebook. In March 2010 convicted sex offender Peter Chapman had admitted to the murder of a 17-year-old girl he had befriended by posing as a teenaged boy on Facebook. Two months later Thomas Mullaney, a 15-year-old boy from Birmingham, hanged himself after being threatened on Facebook.

A Middlesbrough man was gaoled for 14 years after being convicted in September 2013 for tricking children into sending him sexual videos of themselves by pretending to be pop star Justin Bieber. It was reported that a 12-year-old girl had slashed her arms after he had posted indecent photos of her on Facebook. That May a teenager who had been sent a fake bomb by an online stalker who had found her details on Facebook spoke publicly of the dangers of sharing personal information on the website. Five months later Facebook announced plans to relax its rules on privacy settings required for minors, thus – despite the misgivings of internet safety experts – allowing teenagers to share their profiles with strangers rather than just with their online friends.

On 11 March 2014 *The Guardian* reported that child safety groups had warned that the menace of cyberbullying might create a 'lost generation' amidst concerns over the 'mental health of vulnerable teenagers.' The previous October a survey published by the UK's Anti-Bullying Alliance suggested that – so ubiquitous had it become – 55 per cent of young people had come to accept cyberbullying as a part of life. A survey published in April 2014 by the National Association of Schoolmasters Union of Women Teachers showed that 21 per cent of teachers said that they had been verbally abused online by both pupils and parents. Comments had included: 'You are a paedo and your daughter is a whore' and 'My son will fail now because of you.' When later that month a Leeds schoolteacher was stabbed to death by one of her students, the media noted that her killer's Facebook profile contained an image of the Grim Reaper. (Cf. Parsons 2014.)

This tragic and tawdry catalogue of abuses is exhausting but hardly exhaustive. The realm of online friendship increasingly seems a domain of abuse. Terri Apter (2010: 18) has, for example, noted that research has shown that nearly one third of internet users aged between 9 and 19 had received unsolicited sexual comments online.

On 13 April 2010 the BBC reported that Facebook was 'continuing to resist placing a panic button on its pages despite calls to do so by the head of a British child protection agency.' The BBC quoted Richard Allen, Facebook's head of policy in Europe, who commented that the website was one of the 'safest places on the internet.' On 14 April 2010 *The Independent* newspaper reported that Facebook had been criticized by senior British police officers over its refusal to adopt a ubiquitous panic button – adding that the UK's Child Exploitation and Online Protection Centre had received 253 reports relating to paedophile grooming activities on Facebook during the first quarter of 2010. On 11 July 2010 it was reported that Facebook had finally capitulated to campaigners' demands: that Facebook would introduce allow a panic button application, so that children might report abuse to the UK Child Exploitation and Online Protection Centre. The following month it was reported that this panic button had – in its first month of operation in the UK – prompted 211 reports of suspicious behaviour online. One might assume that Facebook's resistance to the imposition of the panic button mechanism had resulted from a concern that the appearance of such an icon would undermine its utopian illusion of a virtual community founded upon an ideal of friendship and openness, an illusion which continues to underpin Facebook's own public image and its defining vision of itself.

The company's similar resistance to calls for more rigorous modes of privacy assurance on the site may similarly relate to its own ideal of itself – a networking platform which excludes malice and criminality and which therefore has no need for privacy or safety concerns. It is unclear whether this apparent naïveté (exposed, for example, by Ron Bowes, the security consultant who in 2010 published the private details of 100 million Facebook users he had harvested from the site) is truly ingenuous. Such cases as that of the Israeli soldier – reported on the BBC on 17 August 2010 – who posted onto Facebook pictures of herself posing with bound and blindfolded Palestinian prisoners seem to suggest that this domain has passed its age of innocence. In August 2010 Google's Eric Schmidt told *The Wall Street Journal* that he did not 'believe society understands what happens when everything is available, knowable and recorded by everyone all the time.' The newspaper added that Schmidt predicted that every young person would in the future be automatically entitled to change their name on reaching adulthood in order to distance themselves from the permanent records of 'youthful hijinks stored on their friends' social media sites.' It seems that the demise of that most simple mode of innocence – the innocence of privacy – may lead towards a more profound loss of moral integrity on a societal scale. We can, in short, be exposed to too much information. We need look no further than the case of Kimberley Swann (who in February 2009 lost her office job after having described her work as boring on Facebook) to witness this.

In April 2012, shortly before the closure of an internationally notorious website which featured pornographic images of its posters' former sexual partners, the site's owner Hunter Moore told the BBC: 'at the end of the day

I feel like I'm just educating people on technology.' It seems increasingly clear that communities of friendship are not the only aims of online social media. (Cf. Rubin 2014.)

Controversies surrounding the photo/video-sharing website Instagram are a case in point. In April 2012 Facebook spent approximately $1 billion on the acquisition of Instagram. Instagram usage grew by 23 per cent during 2013, but as the BBC reported in November that year the site had 'become a marketplace for a wide range of illicit goods. Thousands of pictures showing drugs on offer can be found on Instagram.'

It should, however, be noted at this point – amidst all this anecdotal hype – that the mass media fixations upon these new-media-related transgressions are not entirely innocent in themselves and may of course betray biases grown from anxieties experienced by more traditional media organizations in relation to the applications of emergent media technologies which may threaten to supersede these older forms and institutions. This is a topic to which we will return.

The redetermination of democracy

The Facebook generation seems to found friendship upon a less materially committed ideal than that performed in real-world relationships. This contrasts with attempts by the Aristotelian tradition to base the concept of friendship upon a more ambitious ideal than that generally encountered in the material world, in order to offer our private relationships as models to which our public lives might aspire. Not only does it set a lesser goal for friendship than the metaphysical ideal: it seems to set its target lower than everyday friendship in a non-virtual society. Rather than aiming towards the philosophical heights, it ends up aiming lower than its actual starting point.

Does the commodification of friendship, the exploitation of the name of friendship, this essential shift in its meaning, the loss of this ancient ideal, really make any difference to our lives at all? Only insofar as the traditions of society, politics and democracy are, in western civilization, intimately grounded in this ideal. Friendship represents an inalienable intimacy founded upon mutual transparency, trust, respect, equality and privacy. It is an immanently private condition which affords a model for public, social and political conduct.

The uniqueness and privacy central to this concept of friendship is undermined by the blurring of the public and the private promoted by new media technologies. These technologies are redetermining traditional boundaries between the private and public spheres, and redefining social experience as a public expression of one's own private interactions rather than as an activity negotiated between the protocols of personal relationships and of civic participation.

The traditional notion of friendship has allowed the intimacy, transparency and authenticity of the private sphere to be projected as a model and an ideal for the conduct of relations in the public sphere. Papacharissi (2010: 133) has proposed that 'the privacy of one's being feeds individual authenticity and self-actualization in ways [...] important to enacting behaviours in public.' We might therefore postulate that the increasing publicization (and therefore the destabilization) of formerly private and intimate modes of interaction may threaten to undermine historically entrenched traditions of public, civic and political relations – relations which have defined such structures as those which underpin contemporary democracies.

So much, then, for the promises of democratic dialogue advanced by these interactive technologies. New media have promised to promote the expression and interaction of individuals, and yet they also seem to assimilate and transform difference into a seamless uniformity. The social networking site's cult of the individual is one which standardizes each individual into just another virtual narcissist.

Is this a process of liberation or an entrenchment of economic exploitation and socio-political control in the guise of liberation? And might such liberation constitute a process of revitalized democratization, or commercial hedonism or cynicism, or anarchic individualism, nihilism or narcissism: a simultaneous process of civic fragmentation and cultural homogenization, heralding a state of uniformity without unity? Is this shift truly egalitarian or might it merely represent a rebalancing of hierarchies in favour of a depthless populism?

So we return to this volume's perennial question: are we empowered by these new technologies, or do we just feel more powerful? Do we, for example, actually *know* more, or do we just *think* we do?

CHAPTER 6

Public Knowledge

Another (not entirely unrelated) question: is there any authenticity without authority? Is a structure of authority necessary for the recognition and validation of authenticity?

If history now appears to have become a depthless, insubstantial image of itself, what then is the status of that commodity which we call *knowledge*? Where does knowledge reside – and can we still learn from it?

Wikipedia

Digital democracy does not appear to have significantly increased civic or political participation or democratic accountability. Interactive modes of popular entertainment – from video games to reality television – have not substantially enhanced the agency of their audiences. Social networking websites have not as yet engendered a dialogical public sphere. What then of the area upon which the revolutionary potential of information technologies might be expected to be most appropriately focused: information itself? Have these technologies deconstructed the hierarchies traditionally associated with access to – and the generation of – knowledge?

In this context, it seems inevitable that one examines the internet's – and indeed the world's – foremost source of knowledge, that global phenomenon known as Wikipedia: 'the largest and most popular encyclopedia in the world' (Anderson 2011: 12). It is difficult to imagine a more pervasive information source. By 2014 Wikipedia was able to announce that it boasted '30 million articles in 287 languages.' Up to 40 million words have been added each month to the English language edition alone (cf. Ayers et al. 2008: 4; Lange et al. 2010: 5). Cohen (2014) has reported that Wikipedia is the world's fifth most popular website with '18 billion page views and nearly 500 million unique visitors a month.'

Launched on 15 January 2001, Wikipedia has revolutionized not only our access to knowledge but also our notion of what knowledge is. Despite its avowed requirement to ground all of its content upon authoritative sources – couched in its own imperative that 'all material in Wikipedia articles must be attributable to a reliable published source' – Wikipedia has become notorious for its inaccuracy, while its popularity has simultaneously boomed. This apparent paradox parallels the typical academic response to the site – a combination of public disapprobation and extensive private use. It may not simply be the case that, while one recognizes the platform's unreliability, one enjoys its convenience; the ubiquity of Wikipedia has made its unreliability almost an irrelevance. Insofar as the recognition of discourse as knowledge is an act of consensus (or a process of passive acceptance mediated by structures of power), the content of Wikipedia has become 'knowledge' precisely and only by virtue of its presence on that platform. At the same time, however, our acknowledgement of the content of Wikipedia as knowledge is unprecedentedly provisional: the authoritative status of the site remains ambivalent, and users of Wikipedia appear to be more aware than, say, visitors to the *BBC News* website of the unreliability of the knowledge that this platform supplies them, while also being paradoxically willing to accept Wikipedia as their primary (or, for some, only) source of such knowledge. As such, Wikipedia is posited as a definitive authority which lacks authority; and the fact that for the majority of its users this does not appear to be an insurmountable problem suggests that Wikipedia may be promoting (or may at least be symptomatic of) an increasing mistrust of epistemic authority.

In a blog entry of February 2006 the renowned cybersceptic Andrew Keen wrote that 'the truth about Wikipedia, the unintended consequences of its radical democratization of knowledge, is that it turns everyone into kids.' It remains unclear whether, as some might suggest, this destabilization of received knowledge, the consequent disruption to the authority of professional expertise and the growth of what Keen (2008) famously described as the cult of the amateur herald the beginning of the end of polite civilization as we know it – or whether, as the followers of that epistemological historian Michael Foucault might suppose, this phenomenon might prompt a dynamic restructuration of the ways in which discourse mediates knowledge and power, and, in doing so, promote a perspective which recognizes the inadequacy and absurdity of that which we call 'knowledge' and which therefore subverts the authority of knowledge and of all authority which is based upon knowledge (which is all authority). The former perspective predicts a future of cynical distrust; the latter anticipates a state of sceptical mistrust. This is perhaps the difference between anarchy or demagogy or autocracy on the one hand (depending on what might result from the feared demise of representational politics), and democracy on the other.

Robert Cummings (2009: 1) has distinguished between those perspectives which denounce 'the evils of Wikipedia with the inevitable accompanying lament on the fall of standards and credibility associated with the waning of print culture' and the notion of Wikipedia as 'a harbinger of a new way of writing – and of working.' While some might not share Cummings's enthusiasm for this emergent mode of cultural production and reproduction, it is difficult to suggest other than that Wikipedia and its ilk are here to stay, and that their influence upon culture and society will continue to be ubiquitous and intense – and that, as we can neither deny nor reverse their impact, it is urgent that we engage with these forms to understand the nature of that impact. This chapter therefore specifically explores Wikipedia in its own terms: it examines how Wikipedia defines and determines itself, and thereby investigates the modes of culture and society which Wikipedia might represent and propagate: a system which is, as O'Sullivan (2009: 1) has suggested, 'alien to our cultural traditions [...] operated on the whole by ordinary people rather than academics or professional writers, whereas we live in a society in which the authority to pronounce on matters of fact of any complexity is regarded as the province of experts.'

John Broughton (2008: xv) has noted that 'Wikipedia has never lacked skeptics. Why expect quality articles if everyone – the university professor and the 12-year-old middle school student – has equal editing rights?'

Broughton (2008: xvi) has however pointed out in defence of the site that 'Wikipedia has a large number of rules about its process [...] there are a few editors with special authority to enforce the rules.' While Broughton argues that Wikipedia functions by consensus, Wikipedia is then in many respects an elitist hierarchy not so different from those in the traditional lifeworld: it does not break away from this problematic model so much as it merely adds other problems, problems of authenticity and formal accountability, to that model.

Reagle (2010: 152) has compared the revolutionary potential of Denis Diderot's *Encyclopaedia* to that of Wikipedia: 'Much as the *Encyclopédie* challenged the authority of the church and state and recognized the merit of the ordinary artisan [...] Wikipedia is said to favour mediocrity over expertise.' O'Sullivan (2009: 23) has pointed out that Wikipedia resembles the Library of Alexandria as a project aspiring to accumulate all of the knowledge in the world. Does Wikipedia therefore represent the 'new Alexandria' imagined by the likes of Koskinen (2007: 117) and Tiffin and Rajasingham (2003: 26) – or has the information age merely reinforced the relationship between knowledge and power within an entrenched knowledge economy?

Power to the people

Wikipedia reminds its contributors not to 'worry about making mistakes' on the grounds that 'no damage is irreparable.' Wikipedia has reassured us that 'on average, only a few minutes lie between a blatantly bad or harmful edit, and some editor noticing and acting on it.' It is clearly difficult to see what harm could be done in only a few minutes by a website which is the 'largest and most popular general reference work on the internet' and which offers 30 million articles, produced by about 100,000 active contributors, on subjects ranging from advice on cooking custard through the varying ages of sexual consent in different countries and the mechanics of the jet engine to brain surgery.

There are two mutually exclusive schools of thought on the subject of the contribution to the sum total of human knowledge of the free online encyclopaedia whose content is entirely generated by its dilettante users. One – which we might classify as conservative (one which seeks to maintain entrenched processes for the institutional canonization of knowledge) – would suggest that knowledge is best left in the hands of the professionals: that Wikipedia represents a dilution and a corruption of properly evidenced and provenanced knowledge; for, as Wikipedia has put it in its own description of itself, 'Wikipedia's departure from the expert-driven style of the encyclopedia building mode and the large presence of unacademic content have been noted several times.' Those who hold this position tend to cite the many cases, oft celebrated by the mainstream, media, of Wikipedia's disastrous errors and embarrassments. Some of Wikipedia's more publicly embarrassing incidents include the 2007 case of Ryan Jordan, the Wikipedia editor who passed himself off as a professor of religion at a private university but was in fact a college student who numbered *Catholicism for Dummies* among his primary sources of theological knowledge.

As Dalby (2009: 138) points out, the most serious error made by Ryan Jordan – in a world in which everyone from professors to schoolchildren are deemed to have an equality of voice – was not that he was an amateur but that he pretended to be otherwise: 'to make his pseudonymous identity more weighty and academic than his real persona.' It is when Wikipedians fall back upon an elitism, the institutionally determined hierarchy of knowledge authority which they purport to disavow, that the authenticity and idealism of the project are critically undermined – and its purportedly revolutionary potential is reversed to the extent that it reinforces the entrenched relationships between that-which-is-perceived-as-knowledge and elitist institutional power.

In September 2012 it was reported that the celebrated American author Philip Roth had been unable to convince an Wikipedia administrator of the validity of a correction he wanted made to an entry about one of his novels: he was refused on the grounds that he 'was not a credible source.' So can we really frame this absurdity as an instance of the liberating diminution of authorial power – of entrenched authority – in favour of popular empowerment: is this a case of what Roland Barthes (1977) might have described as the birth of the reader being made possible by the demise of the author?

When Wikipedia erroneously announced the death of the veteran British journalist Alexander Chancellor in 2009 he alluded to Mark Twain's famous response to newspaper reports of his own death by describing the announcement as 'an exaggeration' – and, in doing so, reminded us that advances in media technologies had not necessarily enhanced the accuracy of public reportage in the 112 years since Twain's remark. Less amused was the veteran American journalist John Seigenthaler, who described as 'internet character assassination' the Wikipedian claims that he had been a suspect in the assassinations of President John F. Kennedy and of his brother Bobby Kennedy.

In April 2007 Larry Sanger, one of Wikipedia's founding editors, raised questions as to the site's integrity, prompting *The Guardian* to observe that 'universities have long questioned the reliability of information posted and edited on Wikipedia.' Those universities which eschew the use of Wikipedia of course have vested interests in undermining the credibility of any organization which competes with them in the dissemination of knowledge and information – and whose very existence calls into question the authority of their institutional professionalism.

Reproductive systems

Wikipedia declares that all articles 'must be written from a neutral point of view' and yet it fails to observe that the notion of the authoritative primacy of neutrality and objectivity offers a rationale for entrenched institutional authority: the academic scientificism which denounces subjectivity, and which therefore maintains the myth of the possible existence of objective truth (as if there could ever be a truth independent of subjective perception and interpretation).

In December 2005 the scientific journal *Nature* published a paper which employed 42 'expert reviewers' to compare the website of *Encyclopaedia*

Britannica with Wikipedia. *Nature*'s experts had discovered errors at a rate of about three per *Britannica* article and about four per Wikipedia article. When *Britannica* queried *Nature*'s findings, the journal responded that its 'comparison was unbiased' and that the journal would 'stand by the story' (*Nature* 2006: 582). The approach taken by *Nature* might be seen as going some way towards undermining the institutional position on Wikipedia (Wikipedia is only a third more inaccurate than *Britannica*); yet, insofar as *Nature*'s study depended upon 'expert reviewers' for its comparison, its perspective remained that of the established and accredited professional: its strategy did not call into question the assumption that what we perceive as knowledge might not require the authority of institutional provenance at all; nor did it interrogate the very notion of objective accuracy itself. Indeed, in that it was questioning the accuracy of a popular reference work (*Britannica*) rather than that of a peer-reviewed academic journal, one might suppose that the primary function of *Nature*'s article was not to laud Wikipedia but to discredit *Britannica* in a manner which sought to reinforce the significance of its own academic authority.

A less self-serving celebration of Wikipedia's contribution to knowledge might draw upon Michel Foucault's idea that the categorization of discourse as truth is dependent upon that discourse's relations with the dominant power structures of the culture and society from which it originates and in which it functions: 'it is in discourse that knowledge and power are joined together' (Foucault 1998: 100). That narrative which is acknowledged as knowledge is the one which best suits the hegemonic perspective. A genealogy of knowledge would demonstrate that there has never been any absolute scientific, philosophical or religious truth, but that at any given moment in history pieces of narrative discourse are validated – by academic, cultural, social, political or religious power – as absolute and incontrovertible truths, only to be re-evaluated and discredited in future years as power relations shift towards new models. This is necessarily a dynamic process, for, as Foucault (1998: 99) argues, 'relations of power-knowledge are not static forms of distribution, they are matrices of transformations.'

Natalie Fenton (2013: xi) has reminded us that 'because the relationship between democracy and media is so complex and contingent it is also never fixed and constantly open to contestation.' Fenton has added that

media and democracies are 'not homogenous, static entities' but, in their
ever-changing ways, 'shape the space where power is competed for.' From
this perspective Wikipedia may be seen as able to redefine those areas of
discourse recognized as knowledge with the authority not of fusty institutions of power but of the will of the people. This transformational process
might seem at first sight to be terribly democratic.

Yet this does not necessarily suggest the possibility of individual
empowerment. As Bourdieu (1977: 184) reminds us, society has moved
away from a structure in which power relations are determined by subjective individuals rather than by objectified institutions: away, that is, from
'social universes in which relations of domination are made, unmade,
and remade in and by the interactions between persons' into a realm of
'social formations [...] mediated by objective, institutionalized mechanisms.'
Thus, Bourdieu continues, 'relations of domination have the opacity and
permanence of things and escape the grasp of individual consciousness
and power.' These formations tend 'to reproduce the objective structures
of which they are the product' (Bourdieu 1977: 72). The complicity which
such a structuration requires, although sustained by an illusion of individual
agency, does not require, support or allow such agency, and may not in itself
represent an entirely conscious complicity: its agents (agents as puppets or
drones, agents without autonomous agency) may remain unaware of their
own role and of their lack of autonomy – the agent's 'actions and works are
the product of a modus operandi of which he is not the producer and has
no conscious mastery' (Bourdieu 1977: 79). Wikipedia, with its repeated
emphases upon its own principles, processes, structures and hierarchies,
does not appear to reverse this trend.

If Wikipedia were a democracy, then it would be an explicitly hierarchical one: one whose governance is overtly imposed from the top down.
Wikipedia itself explains that it 'has established a bureaucracy of sorts,
including a clear power structure that gives volunteer administrators the
authority to exercise editorial control.' It has added that 'editors in good
standing in the community can run for one of many levels of volunteer
stewardship; this begins with administrator, a group of privileged users
who have the ability to delete pages, lock articles from being changed
in case of vandalism or editorial disputes, and block users from editing.'

Above the administrators there are bureaucrats (who enjoy the 'ability to add or remove admin rights'), the Arbitration Committee (described as 'Wikipedia's supreme court'), the stewards ('the top echelon of technical permissions') and Wikipedia's surviving co-founder Jimmy Wales who has 'several special roles and privileges' which remain unspecified – though Dalby (2009: 10) has observed that Wales 'holds dictatorial powers to ban users, delete pages and erase page histories.'

Wikipedia defines an autocrat as a person 'ruling with unlimited authority' and notes that 'the autocrat needs some kind of power structure to rule.' It is not clear what limits, if any, there are to Mr Wales's power and privileges within his electronic domain, but it is apparent that his position is supported by a rigid structure of power. Interviewed in *The Independent* (Burrell 2010), Wales – a fan of the UK's House of Lords – has compared Wikipedia to that upper house: 'He talks of the lack of a written constitution, refers to the website's highest body (its arbitration committee), and notes that if you become an administrator in Wikipedia, you are pretty much in for life.' Wales has added: 'I'm working as hard as I can to make it as ceremonial as possible ... a constitutional monarchy where I have certain reserved powers.' Yet alongside these entrenched power structures the site continues to maintain its claims as to its potential to liberate. On 1 May 2013, for example, Jimmy Wales told the BBC that Wikipedia 'plays a vital democratic role in allowing ordinary people to become informed in a way that would never have been possible before.'

If Wikipedia has developed a pyramidal hierarchy, then Axel Bruns (2008: 141) observes Jimmy Wales poised 'at the very top of this pyramid [...] described by some Wikipedians as the site's god-king.' Bruns (2008: 141) has expressed concerns in relation to the website's increasingly entrenched hierarchization: he argues that Wikipedia's principles of open participation and 'heterarchical governance' may be threatened if 'current trends towards the development of more permanent administrative structures continue.' The core pluralist principles of Web 2.0 may be unsustainable when heterarchy is reformulated as hierarchy to avoid its otherwise inevitable collapse into anarchy.

Reagle (2010: 119) has, however, defended Wales's position: 'While a founding leadership role has some semblance of authoritarianism to it [...] it is entirely contingent: a dissatisfied community [...] can always leave.' Reagle's argument, similar to one advanced in 2010 by Google's Eric Schmidt – that unhappy users can vote with their virtual feet – hardly takes into account the virtual monopolies these organizations hold. Reagle fails to recognize the influence of the socio-cultural *habitus*, nor indeed does he acknowledge the power of the *status quo*, a power enforced by the entrenchment of possession and the inertia of habituation. Reagle's argument, were it true, should surely also be true for any political situation, online or off. (Tell that then to the citizens of oppressive regimes.) There is – as Christian Fuchs (2008: 320) suggests – 'a permanent, dynamic self-organization process in which Wikipedia structures and Wikipedians' actions mutually produce each other' – just as, in Bourdieu's terms (1977: 72), the agents of the habitus tend to 'reproduce the objective structures of which they are the product.' This is not then a matter of individual self-determination (although these structures offer the illusion of that) but of the perpetuation of objective systems.

It is fortunate then that Wales maintains his status as – in the words of both Reagle (2010: 177) and Bruns (2008: 147) – a 'benevolent dictator' (but then how many dictators were openly described by their subjects as malevolent while still in power?). Wales's position is grounded upon a series of 'policies and guidelines' which support the site's 'five pillars' (its statutes of principle) and which determine the site's content and development. There is a formal power structure here, albeit one which actively invites participation. If its development does in fact represent a cultural power shift, then Wikipedia does not convey intellectual authority into the domain of the masses so much as it establishes an alternative epistemological elite. As it announces itself, 'Wikipedia is not an experiment in democracy' – 'Wikipedia is not a democracy.'

Wikipedia's own definition of democracy has noted that 'equality and freedom have been identified as important characteristics of democracy since ancient times. These principles are reflected in all citizens being equal before the law and having equal access to power.' Wikipedia may be *free* in the sense that it is free to access; but its hierarchy of roles and rights and its prescriptive principles are not designed to promote freedom and equality within the website's own structures. The Wikipedian enjoys the illusion of an active participation in the creation of knowledge; and yet, as the site has itself emphasized, 'Wikipedia is not a publisher of original thought' and 'does not publish original research.' Within these limitations, the Wikipedian acts only as an aggregator and disseminator of received knowledge, a gofer to external authorities and internal interests – or, as Samuel Johnson once described the lexicographer, a 'harmless drudge'.

Wikipedia has proudly announced that 'when *Time* magazine recognized You as its Person of the Year for 2006, acknowledging the accelerating success of online collaboration and interaction by millions of users around the world, it cited Wikipedia as one of several examples of Web 2.0 services.' The fact that *Time* magazine chose 'the millions of anonymous contributors of user-generated content' as its Person of the Year reinforces the illusion of significance, agency and empowerment afforded to their users by the likes of Wikipedia and Facebook – an illusion which may sublimate the desire for real-world participation and allow those users to continue to function as indistinguishable cogs in the post-industrial economy's knowledge machine. Wikipedia has itself explained that 'no editor owns any article; all of your contributions can and will be mercilessly edited and redistributed.' Wikipedia alienates its workforce from their labour and appropriates the products thereof without even any of the usual material compensations which industrial capitalism once offered. And it expects to be applauded for it.

Wikipedia has described itself as a 'free, web-based, collaborative, multilingual encyclopedia' and it seems significant that it prioritizes the fact that it is *free*. Knowledge may, as Bacon suggested, be power; but it is specifically the ownership of knowledge – the authority of the source of knowledge, the control of the flow of knowledge – which determines the possession of power. Those who own knowledge and control its means of production, mediation and dissemination are, in these terms, rather more powerful than those who merely consume (or for that matter reproduce) it. To be a knowledge-consumer is not a form of empowerment in itself. Wikipedia can supply mediated knowledge for free because that is not the same as giving away power; the power resides in being the knowledge-provider, in being part of (and specifically being at the top of) that organization's hierarchy.

Wikipedia favours and empowers those who have the economic, educational and technological advantages required to access and exploit the possibilities of the internet; it also promotes the interests and perspectives of those who have the leisure time and the intellectual capital to fill its pages and rise through its ranks. Wikipedia propagates the world-view of a structured and regulated elite, an arbitrary quasi-meritocracy no more open or progressive than that already known in the realms of intellectual, political and economic capital, and offers this elite the semblance of an extraordinary degree of influence, authority and power.

It should also be noted that the celebrated openness of Wikipedia does not exclusively advantage amateur influence. Concerns have, for example, been raised in relation to the potential for corporate and governmental interference in the workings of Wikipedia. There are clearly opportunities here for established commercial and political powers to deploy their significant resources to influence public knowledge and perceptions of their activities. There have been various reported incidents of such interested parties amending Wikipedia entries to benefit their own positions. Stewart (2014: 139), for example, recalls how *The New York Times* was able to 'manipulate' news coverage of the kidnapping of one of their reporters by repeatedly deleting reports of the incident from Wikipedia. Other cases may seem less benevolent. In 2007 it was reported that tracking software had identified CIA changes to the Wikipedia entries on Ronald Reagan and Richard Nixon, as well as self-serving revisions emanating from the British Labour Party and the Vatican. In one instance, a company which built electronic voting machines had removed concerns as to the security of its products; in another, Republican Party computers had proven the source of an alteration whereby the American 'occupation' of Iraq had been transformed into an act of 'liberation'. One of the more extraordinary of such cases appeared in April 2014 when claims emerged that offensive revisions to the Wikipedia page on the 1989 Hillsborough disaster (which had resulted in the deaths of 96 football fans) had originated from UK government computers. These amendments had included the addition of the sentences 'Blame Liverpool fans' and 'This is a shit hole.' Later that month it emerged that further inappropriate edits had emanated from British government computers, including instances of homophobic abuse, the assertion that 'all Muslims are terrorists' and the replacement of the entire entry on Tony Blair with the words 'he should be assassinated.' These revisions also included the claim that the popular TV presenter Desmond Lynam had been 'killed by a giant snowball.'

The destabilization of knowledge

Wikipedia's continuing refusal to follow the other global giants of Web 2.0 in their pursuit of financial gain and in their development of commercial interests (Wikipedia refuses to take advertising: advertising, is says, 'doesn't belong here') suggests that the site has not perhaps entirely abandoned the radical potential of its founding ideals. Could Wikipedia's destabilization of the concept of knowledge then stimulate the development of a problematized popular understanding of knowledge, of its limitations and its uses, which might advance the evolution of a more democratic perspective upon authority *per se*? Democracy is based upon trust, but that trust cannot be blind; democracy requires that authority be challenged (and therefore be mistrusted) so that it might eventually and provisionally be accepted (and therefore not be distrusted): in order to avoid a collapse into cynicism, democracy requires the maintenance of scepticism, and that scepticism might originate at the level of epistemology.

Wikipedia is an ever-changing organism, and, as such, challenges from the outset the academic's desire to cite the hard evidence of permanently fixed texts. Knowledge, we like to think, should be as solid as the bricks and mortar which house its ancient institutions; the revelation that knowledge is fluid may be somewhat distressing to those who have founded their entire careers upon such ambiguous materials. Yet this revelation is key to any notion of widening participation in the processes of knowledge, and therefore in the processes of power. However, although Wikipedia's inherent mutability may offer to subvert any fixed, formal and deferential notion of the authority of established knowledge, its avowed mission – that of the public encyclopaedia – necessarily relies upon and perpetuates such a sense of authority. In its attempts to be encyclopaedic – to be comprehensive, objective and neutral – it cites established knowledge and its own self-regulating hierarchies (not dissimilar to those of academia) as the provenance of its authority; but it is through its failure to become a source of knowledge as immutably authoritative as those it cites (simultaneous with its ability to become a source more influential than those traditional sources) that Wikipedia prompts a radical shift in popular perceptions of knowledge and its authority.

Yet when we examine Wikipedia's own perspectives on knowledge itself, we return to that intrinsic conservatism which undermines the portal's revolutionary promise. Wikipedia's entry on 'knowledge' has *disambiguated* its primary notion of knowledge as 'the possession of information' – distinguishing this concept from 'the rigorous geographical training obligatory for London taxi drivers' and from – amongst other things – a Jamaican reggae group, three television channels and a series of children's books, all of the same name. Knowledge is about possession – that is its first and most prevalent sense.

The Wikipedia entry on 'knowledge' quotes the *Oxford English Dictionary*'s definition and cites Plato's formulation of knowledge as 'justified true belief' – before going on to reference Aristotle, Russell, Rorty and Wittgenstein, amongst others, and supposing that 'the definition of knowledge is a matter of on-going debate among philosophers in the field of epistemology.' This debate is clearly the province of qualified philosophers rather than a matter of concern for the public at large.

In spring 2014 the Wikipedia entry on knowledge included 624 words on the communication of knowledge, 462 words on scientific knowledge, 404 words on religious knowledge and only 108 words on the partiality and incompleteness of knowledge. This emphasis upon the stability, certainty and authority of knowledge (rather than upon its ephemerality) is echoed in the illustrations which have accompanied the article: a portrait of Sir Francis Bacon, a Greek statue personifying knowledge in amongst the ruins of the ancient Library of Celsus in Ephesus, and a mural by Robert Reid dating from 1896 and depicting an anthropomorphic idealization of *Knowledge* – a young woman holding a large book. Reid's mural adorns a wall of the Library of Congress in Washington D. C. above a caption which announces that 'knowledge is the wing wherewith we fly to heaven.' The message of Wikipedia's text and its illustrations seems to be that knowledge is for the most part reputable and reliable; it is the product of professional scientists and seers; its discussion is the province of accredited philosophers and academic authorities; its place – indeed, it source – is the library; it is the immutable and absolute key to the divine, to enlightenment and to power (but only via these figures, symbols and institutions of authority).

We may however note (as we may discover from Wikipedia's own entry on the painter Robert Reid) that in the north corridor of the Library of Congress's second floor Reid's representation of *Knowledge* is accompanied by a similarly idealized embodiment of *Wisdom* – beneath which stand the words 'knowledge comes but wisdom lingers.' This maxim (along with the derelict state of the antique Library of Celsus) may serve to remind us that knowledge is as ephemeral as the power which (for Bacon) it promotes and which (for Foucault) also promotes it. If Wikipedia is the new Alexandria, then we should not forget what happened there, nor indeed how little we actually know about that library's fate: as Wikipedia itself pertinently observers, 'sources differ on who is responsible for the destruction and when it occurred.'

Wikipedia seeks to create the conditions for a homogeneous and authoritative consensus of knowledge, a new canon and world-view constructed by – and initiating – an emergent epistemological elite, a prescribed and regulated hierarchy founded upon, and usurping, the authority of earlier intellectual hegemonies. However, in doing so it unwittingly exposes the fragility of, and thereby undermines the authority of, the concept of established knowledge – the very myth from which Wikipedia endeavours to develop its own power. Wikipedia is not democratic (as it says it is not); its mission is as authoritative as those structures it seeks to supersede; and yet its failure to achieve its mission opens up possibilities for the promotion of democracy.

Insofar as its contributors lack the trappings and symbols of academic power (the reassuring regalia of dress, title, tenure, ceremony, architecture, lecture theatre and print publication) which tend to veil the incoherence of academic knowledge, the absurdity of Wikipedia's operations (conspicuous in their amateurishness) may be seen as subverting the platform's attempts at epistemic authority. But at the same time Wikipedia has become synonymous with knowledge: it has become the post-industrial world's most influential source of information. It views knowledge as 'the possession of information' – knowledge is its possession; and the general recognition of Wikipedia's primacy in the ownership of knowledge is crucial to its authority and power. Wikipedia *is* knowledge: it is the domain of knowledge; knowledge is its domain.

Wikipedia is also patently and famously unreliable and ephemeral. Try as it might to establish the authority of knowledge *per se* (in order to establish its own authority) Wikipedia inevitably calls the authority of knowledge, of all knowledge, into question, and thereby undermines an unquestioning acceptance of authority itself. As Foucault (1998: 101) has suggested, while 'discourse transmits and produces power; it reinforces it, but it also undermines and exposes it, renders it fragile and makes it possible to thwart it.' This thwarting of the structures of power is not necessarily an advertent process. Despite itself, and despite its hierarchical and even autocratic tendencies, and as an unintended consequence of its own failings and contradictions, Wikipedia may thereby offer an opportunity for the development of democracy, in that the tendency to challenge – and thus to begin to *thwart* – the structures and processes of power is crucial to active and participatory citizenship in a pluralist and democratic society. But it may only be when Wikipedia and its users become aware of this potential (a potential which requires an awareness of the fundamental incoherence – and thereby of the fundamental value – of knowledge) that this phenomenon might manage to defer the slide into philistinism or that dictatorship of ignorance which many fear Wikipedia now represents.

The knowledge economy

If Wikipedia has come to represent the sum total of human knowledge, then Wikipedia offers a model or a map of the world which increasingly subsumes the world itself. As such, it has come to resemble the fantasy realm of Tlön, Borges's world imagined at first as a theoretical model by intellectuals which eventually becomes the model on which the material world bases itself (Borges 1970) – or Baudrillard's notion of the map of the world which becomes the world (Baudrillard 1994). In short, the simulacrum has taken over: but this is a simulacrum constructed not by Borges's academics but by a section (admittedly a privileged section) of the general population. What kind of world, then, would such a project construct? Is this new media utopia to be idealistic or pragmatic, democratic or demagogic, anarchic or autocratic, egalitarian or elitist, meritocratic or vulgarian?

Wikipedia certainly endeavours to be pluralist: 'we strive for articles that advocate no single point of view. Sometimes this requires representing multiple points of view, presenting each point of view accurately and in context, and not presenting any point of view as the truth or the best view.' Wikipedia struggles towards an ideal not dissimilar to John Fiske's best-practice model for journalism. Fiske (1987: 307–308) has argued that in a progressive democracy, news should 'nominate *all* its voices' – in other words, it should name all of its sources and therefore never present any of its perspectives as ideologically neutral. Wikipedia has offered a similar strategy:

> Values or opinions must not be written as if they were in Wikipedia's voice. When we want to present an opinion, we do so factually by attributing the opinion in the text to a person, organization, group of persons, or percentage of persons, and state as a fact that they have this opinion, citing a reliable source for the fact that the person, organization, group or percentage of persons holds the particular opinion.

Wikipedia's willingness to invoke a plurality of voices invites the inclusion of those voices which criticize Wikipedia itself. As of summer 2010, while the platform's entry on the topic of 'Wikipedia' accounted for just under 12.5 thousand words, its section on 'Criticism of Wikipedia' ran to over 14.5 thousand. Two years later, however, while the former page had grown to over 17 thousand words, the latter page appeared to have disappeared – a search for 'criticism of Wikipedia' redirected the browser back to the former page. Two years further on, the page had returned, but now ran to only about eight thousand words. By this time Wikipedia's page about Wikipedia was more than 21 thousand words long.

Despite its varying emphases, the site's openness to criticism would appear to move Wikipedia in the direction of Fredric Jameson's notion of the meta-utopian (Jameson 2005), that mode of utopianism which inscribes its own counter-argument within itself. The site's section on 'Why Wikipedia is not so great' asserts, for example, that 'Wikipedia *is* a bureaucracy, full of rules described as policies and guidelines with a hierarchy aimed at enforcing these.' Might Wikipedia therefore aspire to the pluralist, dynamic criteria for what Tom Moylan (1986: 213) has described as a 'self-critical utopian discourse'?

For as long as Wikipedia draws attention to its own inadequacies and innate contradictions, then it might be seen as having chosen the path of dialogism over demagoguery. This is not, of course, so much an end *per se* as it is an ongoing and dynamic process: one whose participants move continually towards a consciousness of the absurdity and the inevitability of the relationship between knowledge and power, and yet one which may be swiftly undermined by the re-imposition of rigid hierarchies.

The danger is, of course, that Wikipedia becomes a self-sustaining and self-censoring autocracy based upon an elitist hierarchy of knowledge-authority which not only magnifies the disparities and distortions of informational institutions in the physical world, but which also lacks the historical provenance and public accountability upon which those institutions were founded – and that therefore the value of the information it provides is only as a matter of symbolic exchange which underpins the socio-political positioning of its elite. Dalby (2009: 177) has pointed out Wikipedia's tendency to offer as its sources websites which are themselves only mirrors of Wikipedia (sites which have already copied their content from Wikipedia); in these cases, Wikipedia seems a depthless, unprovenanced and unquestionable simulacrum of knowledge, a reflection without an original object – a symptom of the fetishization of knowledge as an economic commodity, rather than an opportunity to democratize not only access to knowledge but also the production, interpretation and valorization of knowledge.

Insofar as Wikipedia bases itself upon, and therefore compounds, the structures and principles of institutionalized knowledge, it perpetuates and extends an age-old trade in knowledge as socio-economic capital. The citizen becomes a knowledge-consumer rather than a knowledge-producer, in that the individual's contribution to knowledge (their interpretation and research) is denounced as invalid, as a corruption of the purity of fetishized information. In these terms, Wikipedia's logo – its incomplete jigsaw globe – might recall the half-assembled Death Star in *Return of the Jedi*: a mechanism which, once complete, will devastate all within its path.

CHAPTER 7

The Twitterati

Andrew Keen has argued against the 'real political reactionaries' in the emergent elite of the digital media industry, the new 'antiestablishment establishment' (Keen 2008: xviii, 13). He berates the internet's 'souring' of civic discourse and asserts that 'the decline of the quality and reliability of the information we receive [is] distorting, if not outrightly corrupting, our national civic conversation' (Keen 2008: 15, 27). This represents what Keen (2008: 54) has dubbed 'the degeneration of democracy into the rule of the mob and the rumor mill.'

There have been a number of high-profile cases in which social media have become the vehicles for such mob mentalities. Ivor Gaber (2010) – in reference to the use of social networking technologies in the 2009 campaign against *Daily Mail* journalist Jan Moir's attack upon the late Stephen Gately – has spoken of the emergence of a 'Twitter mob rule – a Twitter dictatorship.' Or, as the Conservative MP Louise Mensch told *The Independent* newspaper in May 2012, 'if you want to see the worst of humanity, look on Twitter.'

Twitter Twatter

In the second episode of the 2014 BBC comedy series *W1A*, a PR consultant sets up a Twitter account for her client – and starts tweeting on his behalf – in order to establish his 'cultural capital'. Things of course go terribly awry. This comedic misadventure mirrors the real-world consequences both hilarious and disastrous that have so often and so very publicly accompanied attempts to exploit this particular social media platform to bolster cultural, social or political capital. Where Facebook may be critiqued for its emphasis on the economic value it offers its users under the guise of friendship (value, that is, instead of *values*), Twitter appears to have become notorious for its much-publicized failures to provide even that.

The profile and influence of Twitter has grown exponentially (and, some might say, unsustainably) in recent years. In May 2012 Twitter topped ten million users in the UK alone. On 4 October 2013 the BBC reported that the company was planning to raise a billion dollars in its first stock market flotation. The report added that, although the website boasted 218 monthly users, it had incurred losses of $69 million in the first half of 2013 and had 'never made a profit.' A month later (on 5 November) the *Financial Times* reported that the company had increased its planned share price by 25 per cent in response to 'huge investor demand' – and was thus estimating its own value as in excess of 17 billion dollars. Two days later the BBC noted that the final share price announced had valued the company more than 18 billion dollars, and added that 'its losses for the third quarter of 2013 increased to $64.6m, from $21.6m a year earlier.' Later that same day the BBC noted that Twitter's share price had soared by 73 per cent in the first few minutes of trading on the New York Stock Exchange, and that the company's value was therefore considered to be a little over 31 billion dollars. Twitter reported losses of $645m in 2013, and the publication of these figures in early 2014 resulted in a 23 per cent fall in its share price. The company's total value had been reduced by 6.5 billion dollars. Twitter's report for the first quarter of 2014 boasted that the platform's number of monthly users had risen to 255 million, but posted an overall loss of $132 million and noted that its quarterly recruitment of new users had dropped to 14 million from 19 million during the same period the previous year. It also recorded an 8 per cent decrease in total user engagement with the site.

There is clearly something somewhat absurd in all this. In the first volume of his account of *Capital*, Karl Marx (1976) famously recounted the development of exchange-value (how much people are willing to pay for something) as opposed to use-value (how much it is actually worth in terms of the benefits it offers), and we may observe that the more extreme contradictions between these relational and innate forms of value have often exposed and exploded material inequities and absurdities within capitalist economies.

One recalls the dot com bubble of the late 1990s – the bubble which finally burst in March 2000. More recently, we have witnessed the fluctuations in the worth of that most virtual of currencies, the bitcoin – a currency described by *The Economist* in November 2013 as requiring a greater degree of 'consensual hallucination' than most. At the start of 2013 the value of one bitcoin was about £7 – by November that year it had risen to around £750. That month James Howells hit the headlines when he realized that the hoard of bitcoins which he had acquired for next to nothing some four years earlier was by then worth more than £4 million – but was unfortunately languishing in the memory of his old discarded laptop somewhere within a landfill in Newport, South Wales. Bitcoin had become something of a 'magnet for speculators' which – according to Rupert Jones (2014) – appeared to be on track either 'to change the world by turning e-commerce on its head, or end very messily like a modern-day version of the tulip mania that gripped the Netherlands in the seventeenth century.' In January 2014 eBay banned the sale of bitcoin from its auction and buy-it-now sites. The following month MtGox – once the world's largest bitcoin exchange – 'disappeared from the internet with many millions of dollars of customer deposits' (*The Guardian*, 25 February 2014). The next month the Chinese government announced that banks would be obliged to close any accounts operated by bitcoin exchanges by the middle of the following month – by which time the value of the so-called cryptocurrency had dropped to less than a third of its November worth. A month on, the *International Business Times* was reporting that the value of the currency was still continuing to plummet. (Cf. Brady 2014.)

The violent fluctuations in the corporate value of Twitter, as of the value of bitcoin, reflect the fact that the illusory nature of value is both underlined and heightened by the virtual nature of these kinds of product. In the case of Twitter, the situation is further exacerbated by the somewhat ambiguous value of the product to its users. If the value of the company is based upon the number of its users (and therefore its potential to generate future income if it can develop any viable mode of significant income generation out of this extensive user-base), then any doubts as to the size and sustainability of that user-base will undermine the stability of that value. In other words, the value of the company may be undermined by: (1) questions as to whether or not registered users in fact use the service at all; (2) questions as to whether or not registered users actually exist at all; and (3) questions as to whether or not those users who are in fact real will remain sufficiently satisfied with the value of the service to continue using the service in the long term.

Uncertainty as to the first of these points was raised in November 2013 when a poll undertaken by Reuters/Ipsos had shown that more than a third of registered Twitter users did not in fact use the platform at all. This seems to be an even greater problem for Twitter than for Facebook, although Facebook also has concerns in this area: in August 2012 it had admitted that nearly nine per cent of its user accounts were illegitimate (duplicate, misclassified or simply fake).

On 11 April 2013 the BBC reported that around half of the pop star Justin Bieber's Twitter followers did not in fact exist: 'out of his 37.3 million followers, only 17.8 million are linked to real accounts.' It is not of course only the followers of the famous but also the Twitter accounts of the famous themselves which might be counterfeit. In June 2009 a Twitter hoaxer posing as Labour Foreign Secretary David Miliband had posted an emotive response to the death of Michael Jackson ('Never has one soared so high and yet dived so low'); three years on, in August 2012, his brother, Labour Party leader Ed Miliband, had joined former footballer Gary Lineker and Martha Lane Fox (the British government's Digital Champion) in congratulating a Twitter user posing as filmmaker Danny Boyle on the success of his opening ceremony for the London Olympic Games. That same month another hoax suggested that champion cyclist Bradley Wiggins had harangued the journalist Piers Morgan on the site. In February 2013 the fast-food giant Burger King announced that its Twitter account had been hacked, after it had started promoting the rival chain McDonald's and in particular their Fish McBites, posting racist remarks and other obscenities, and claiming that its burgers contained the psychoactive drug methylenedioxypyrovalerone.

In April 2011 a Labour Party candidate for election to the Welsh Assembly apologized after declaring on Facebook that he hoped that former Prime Minister Margaret Thatcher would soon die. Two years later, in April 2013, a police officer resigned after writing on Twitter that he hoped that Baroness Thatcher's death that month had been 'painful and degrading.' Other occasions of Thatcher-related online controversy have been less obviously intentional. On 14 November 2009 the *Daily Mirror* had reported that a text message sent by Canadian transport minister John Baird announcing that his cat – also named Thatcher – had died provoked chaos at a diplomatic event in Toronto; three and a half years later the actual death of the former British Prime Minister Margaret Thatcher had prompted a rather less predictable brand of confusion when, as the *Daily Mail* reported on 9 April 2013, 'horrified Cher fans thought their idol was dead after Thatcher critics started #nowthatcherisdead hashtag on Twitter.' The verbal conflation afforded by the hashtag had also resulted in embarrassment for another great diva, the Scottish songstress Susan Boyle, a few months earlier when, as *The Guardian* observed on 22 November 2012, 'the unfortunate choice of hashtag #Susanalbumparty to promote the singer's new album event spawned a wealth of mock invites to the party.'

In July 2009 David Cameron had noted that the trouble with Twitter was its 'instantness' and had joked that, as a result, too many tweets 'might make a twat.' In October 2012, however, Mr Cameron had joined Twitter and had immediately received what a *Guardian* headline described as 'torrents of abuse.' Initial responses to David Cameron's presence on Twitter had included: 'Why are you such a prick?', 'You moon-faced arsehole', 'Briefcase wanker' and 'Piss off you absolute cock-womble.' The Prime Minister was not however entirely discouraged by this experience, and it was reported the following month that his faith in social media had led him to trial what the BBC described as 'a computer app designed to help him run the country' – one which drew trending information from such social media sources as Twitter. A note of concern had been struck by Dominic Campbell, founder of technology consultants FutureGov, who warned the BBC on 8 November 2012 that 'trending topics is hardly a way to run a government.'

Other political figures have also found that Twitter did not live up to its promised enhancement of their cultural capital. In January 2010 council members in Cornwall were reprimanded for sending Twitter messages which mocked other councillors during a council meeting. The following month the Conservative Party chairman had complained when a Labour MP had described the Conservatives as 'scum-sucking pigs' on Twitter. Two months later a Labour Party candidate was banned from standing at the forthcoming election after announcing his penchant for drunkenness, describing the elderly as 'coffin dodgers' and posting a series of obscene rants about fellow politicians on Twitter: calling a well-known colleague in the Labour Party as a 'fucking idiot' and the Speaker of the House of Commons an 'opportunist little twat.' Ironically he had a few days earlier tweeted the observation that 'the biggest gaffes will likely be made by candidates on Twitter.'

At the start of May 2014 a UK Independence Party candidate for local elections was obliged to quit the race after tweeting that Muslims were 'devil's kids', that Pakistan should be 'nuked', that homosexuality was an 'abomination before God' and that the Prime Minister was a 'gay-loving nutcase.' The following day saw the withdrawal of a Conservative election candidate following some similarly Islamophobic and homophobic retweets including a suggestion that Islam was a 'religion of peace and rape.' (Cf. Withnall 2014.)

The most infamous of political Twitter gaffes thus far has involved U. S. congressman Anthony Weiner. On 27 May 2011 Weiner took his first step on the road towards nominative-determinist notoriety when he accidentally posted a lewd selfie onto his Twitter account. In July 2013, while ex-congressman Weiner was mounting an unsuccessful bid for the mayoralty of New York, it further emerged that he had – since the original incident – conducted further online liaisons with a number of other women and had adopted the *nom de sext* Carlos Danger.

Weiner is not the only public figure to have confused the public and private functions of Twitter. In February 2009 Republican congressman Pete Hoekstra had raised security concerns when he had tweeted details of an ongoing trip to Iraq. Five months later another Republican politician, California Governor Arnold Schwarzenegger, caused some consternation when he chose (inexplicably) to wield a two-foot-long knife in a video message posted to his Twitter account. In November 2010 the former Republican Governor of Alaska Sarah Palin was embarrassed at having favourited a tweet which suggested that Barack Obama was a 'Taliban Muslim illegally elected President.'

In October 2011 the UK government's Energy Secretary was obliged to apologize to the Home Secretary after comparing her rhetoric to that of the leader of the UK Independence Party – adding in his inadvertently published tweet to a *Guardian* journalist: 'I do not want my fingerprints on the story.' Two years later, in September 2013, the editor of the BBC's *Newsnight* had to apologize after tweeting that a senior Labour politician who had appeared on the programme had been 'boring'. He had intended the message to be private but had accidentally tweeted it to his 26,000 Twitter followers. (Cf. Stevens 2014.)

Other public figures have felt their Twitter utterances coming back to haunt them. In April 2013 the seventeen-year-old Paris Brown, shortly after her appointment as Britain's first Youth Police and Crime Commissioner, was embarrassed when the media found that she had previously posted such tweets as 'Everyone on *Made in Chelsea* looks like a fucking fag' and 'Fucking hell why are people from Direct Pizza so difficult to talk to!! IT IS CALLED ENGLISH. LEARN IT.' That year the UK press had also repeatedly lambasted British National Party leader Nick Griffin for his conduct on Twitter, including an apparent attempt to associate alleged spousal violence with erotic titillation in relation to the domestic contro-versy surrounding celebrity cook Nigella Lawson, a curious disparagement of 'Glaswegian Fenians' and an attack on the recently deceased Nelson Mandela. In April 2014 a UK Independence Party local election candidate also sparked outrage when he tweeted that British comedian Lenny Henry should emigrate to a 'black country' and compared Islam to the Third Reich.

There is something about Twitter which, perhaps even more than other social media sites, appears to provoke antisocial conduct. The potential anonymity of the transgressor and the invisibility of their victim, the dis-tancing of the abuse and of its impact, and the speed of communication and response (which seems to provoke thoughtless escalations of hostilities): all these conditions appear to be significant factors in the propagation of antisocial behaviours, although they are not of course unique to Twitter. Twitter also tends to attract a younger demographic than such sites as Facebook, and it is notable that those social networking sites favoured by teenaged users and structured in ways which promote the evolution of pack behaviours appear to have prompted higher incidences of such phenomena as deaths by 'neknomination' and bullying-induced suicides. Such issues came to prominence in February 2014 – during which five people were reported to have died as a result of neknominations – and in August 2013 when 14-year-old Hannah Smith killed herself after being bullied on the social networking site ask.fm. Her father was reported by the BBC on 6 August as having written 'on Facebook that he found bullying posts on his daughter's ask.fm page from people telling her to die.'

This seems typical of an emerging generation gap between the users of different social networking services. Alice Jones (2014) has written that, while Facebook appeals cross-generationally, Twitter primarily attracts the young: 'Is your mom on Facebook? Yes. You ask that same question about Twitter, the answer is almost always no.' One might add that although neknominating may have started on Facebook, it represents a predominantly adolescent usage of such social media tools; and that sites such as ask.fm and Twitter – in that they have higher proportions of that demographic among their active users than Facebook – are therefore more likely to witness such behaviours.

One might also, however, point out that, if abusive behaviours and their tragic consequences are more typical of adolescence than, say, of middle age, the prevalence of such behaviour on social sites popular with young people is hardly surprising – and that this is therefore not an issue of technology, but of youth. Indeed the discovery of so many documented cases of victimization on Twitter or ask.fm is precisely because online bullying represents a self-archiving mode of abuse. For the first time in history, we are able relentlessly to witness in black and white the trails of threat and insult which preceded – and which may have contributed to – the tragedy of teenage suicide.

This capacity for the medium to return time and again to the traces of its abuse was underlined in September 2013 when it emerged that a dating advertisement had appeared on Facebook featuring the image of a 17-year-old Canadian girl who had killed herself after another photograph – one allegedly showing her being gang-raped – had been circulated online. The dating ad had featured her picture beneath the slogan 'Find love in

Canada!' In response to her death Nova Scotia had created a special police unit focusing on cyber-bullying. Yet, of course, the abuse continues, as it has always done; social media may or may not make it worse, but what is clear is that through social media these crimes and their victims continue to haunt public consciousness.

This may to some extent explain why there is such extensive media coverage of such abuse and its tragic impact: because the evidence of such abuse is so easily accessible – indeed so unavoidable – online. But does the press coverage of such online abuse reflect the endemic extent of such abuse, or the anxieties of entrenched institutions as to their upstart rivals, the social networking sites, these gauche young Turks? Or might the extent of this coverage more broadly (and more fairly) represent a societal anxiety and ambivalence as to the uncertain impact of these emergent modes of mass communication?

When we commit words or still or moving images to the ether, they seem somehow transient, somehow ethereal. That transience – as we write these names in water – may make us forget how public and how permanent such pronouncements in fact are. There is an illusory privacy which underpins this act of global publication.

If Web 2.0 offers the expression, publication or broadcast of the self, then it also promotes the egocentricity of monologue rather than the possibility of dialogue; and – just as the potential for political policy debate has been said to have been undermined by the era of the media soundbyte – so the proliferation of 140-character messages does not necessarily promote consensus-building conversation, nor indeed does it allow for the face-saving tactics and other politeness strategies which we might favour in offline discourse. This – alongside the speed of response and the rising noise of groupthink, as the individual shouts to be heard above and within the baying pack – may add further to the escalation of any underlying potentials for conflict and abuse.

In August 2012 it was reported that BBC children's TV presenter Helen Skelton had quit Twitter as a result of the personal attacks she had received. In December 2013 it was reported that British Olympic diver Tom Daley had been subject to a slew of homophobic tweets since revealing he was in a same-sex relationship: indeed, as early as September 2012 it had been reported that Mr Daley's receipt of an offensive Twitter message had prompted the UK's Director of Public Prosecutions to announce plans for new rules of social media abuse. In December 2012 the Director of Public Prosecutions had revealed the Crown Prosecution Service had by then addressed more than 50 cases of potentially criminal online posts.

Summer 2013 witnessed a series of Twitter-based threats against a number of female public figures. Towards the end of July it was reported that Caroline Criado-Perez (a campaigner for the inclusion of images of women on UK banknotes) had received rape threats via Twitter. Within a few days this abuse had also been extended to several female journalists, and had been escalated to bomb threats. By the start of August the police were investigating the online the harassment of eight individuals and on 4 August it was reported that even the classical historian Mary Beard had been sent a bomb threat via Twitter.

Mary Beard is perhaps best known to the British public for a number of informative, entertaining and slightly quirky (but entirely inoffensive and uncontroversial) television documentaries about ancient Roman history which she has made for BBC Two. The similarly inoffensive BBC Two programmes of another Mary B – Mary Berry – also that year prompted a certain amount of inexplicably misogynistic online abuse, when one of the finalists in *The Great British Bake Off* was subject to a series of hate-filled slurs. On 22 October 2013 the said finalist, Ruby Tandoh, published a thoughtful and articulate reflection upon her experiences in *The Guardian*: 'I have defended myself against accusations of being a "filthy slag" based solely on me being a woman on a TV screen. If a show as gentle as *Bake Off* can stir up such a sludge of lazy misogyny in the murky waters of the internet, I hate to imagine the full scale of the problem.' (Cf. Corcoran 2014.)

Laura Bates (2013) has asked why Facebook appears to have a 'problem with women' – why 'images that seem to glorify rape and domestic violence keep appearing.' Of course, it is not just Facebook: it may well be that Twitter and other sites have demonstrated more profound and extensive degrees of misogyny along with their other forms of hatred. But it is Facebook that has set the model – both structurally and morally – for the conduct of communication in social media sites; and it seems pertinent to recall that Facebook's precursor and prototype Facemash – established by Mark Zuckerberg while he was studying at Harvard – specifically asked users to 'judge' its unwitting subjects on their looks – 'Who's hotter?' On 19 November 2003 *The Harvard Crimson* newspaper had reported that Zuckerberg had faced complaints from the University's computer services department in relation to his site's 'unauthorized use of on-line facebook photographs.' The site had been accused in this connection of 'violating individual privacy.'

This then was where the template for contemporary social networking sites originated. Bates (2013) has suggested that we might reasonably accept that there is some 'relation between the treatment and perception of women in the real world and the cultural norms promoted by the most used social networking site on the planet.' One might add that the general perception and treatment of *all* people – female and male – is also very probably now at stake.

Terms of contempt

Another breed of Twitter abuse had arisen in November 2012 when – following a BBC *Newsnight* feature which had made what were later revealed as unfounded allegations against an unnamed politician – rumours emerged on Twitter that this politician was the Conservative peer Lord McAlpine. McAlpine took legal action against those who had instigated and perpetuated what he described as a form of 'trial by Twitter.' It seems ironic that when the peer died in January 2012 the BBC reported that the Prime Minister's response had come via that very social networking site which had caused Lord McAlpine such distress: 'David Cameron tweeted his thoughts were with his family.' Responses to Mr Cameron's memorial tweet had included: 'he was an evil posh subhuman reptile', 'he was a paedophile', 'hopefully you're next', 'cunt' and 'fuck off dish face.' About 80 per cent of responses adopted a similar tone. Lessons, it seemed, had not entirely been learnt.

The UK's popular press have in recent years begun to exploit a loophole whereby Twitter breaches of legal protocols can be reported without any risk of action being taken against the newspaper itself for perpetuating such breaches. On 1 December 2012 *The Sun*, for example, published a story pointing out that a man arrested (but not named) by police in relation to a high profile sex abuse scandal had been named on the internet: 'a telly legend quizzed by cops [...] has been named amid a flurry of wild rumours on the internet.' The newspaper went on to identify a particular microblogger who had 'tweeted a name' (its readers with internet access could therefore swiftly discover the identity of the celebrity who had been interviewed by the police without putting the paper itself at any legal risk). In another piece the following day the paper reminded its readers that 'the star remained named on the internet.' It may be noted that just a few weeks earlier the same newspaper had denounced the 'irresponsible internet postings' (9 November) and 'internet smears' (10 November) against Lord McAlpine: 'a crackpot conspiracy theory that has been doing the rounds on the internet' (11 November). 'Many tweeters,' *The Sun* had complained on 26 November (just five days before it promoted its next round of online allegations of celebrity paedophilia), 'have been acting as if they were above the law.'

In September 2013 a British soap star was found not guilty on a series of sexual offence charges. That same month three men had been arrested for posting the name of an alleged victim of the said soap star on social media sites. Six months earlier the police had announced that they were investigating an online hate campaign against the soap star which might have constituted contempt of court insofar as it prejudiced his right to a fair trial.

Between February and April 2013 three more high-profile public figures – a disc jockey, another soap star (from the same soap) and a senior parliamentarian – were also found not guilty of historical charges of rape and indecent assault. On 11 April 2014 the former Director of Public Prosecutions Lord Macdonald warned that the UK's Crown Prosecution Service should in future be careful to avoid 'losing perspective' in relation to such 'historical cases which have garnered a lot of publicity.' It seems increasingly clear that much of this publicity has either originated in or been exponentially escalated by the Chinese whispers and snowballing iterations of online social media.

There might appear then to be certain flashpoints for storms of Twitter abuse: middlebrow BBC Two programming, allegations of paedophilia and, of course, such sporting events as football fixtures which are so intemperate as to pitch West Ham against their age-old rivals Tottenham Hotspur: in November 2013 a West Ham fan was arrested for posting a number of anti-semitic tweets invoking Hitler and the Holocaust, and fifteen days later three more West Ham fans were arrested on similar grounds. In March 2012 a custodial sentence had been handed down to another football fan, Liam Stacey, a student who had posted onto Twitter racially offensive comments about a critically ill footballer.

The Times newspaper's crime editor Sean O'Neill argued (in May 2011) that the outputs of the tweeting classes may pose a threat to the equitable processes of the justice system: they 'could prejudice a trial, put a protected witness's life in danger or cause serious psychological damage to a victim of sexual assault.' One particular case of online social networking which proved prejudicial to the process of justice came to light in June 2011 when a former juror who had contacted a defendant via Facebook (thus causing a trial to collapse) was gaoled for eight months for contempt of court. The social networking site's blurring of traditional boundaries within the justice system (as within other areas of society) was also witnessed in March 2009 when prison officer Nathan Singh was dismissed from the service after befriending criminals on Facebook – and in spring 2012 when a number of people were arrested after posting onto Twitter the name of the nineteen-year-old victim of a rape by a Welsh footballer. In July 2013 a juror in a case of underage child abuse was gaoled after writing on Facebook of his intention to 'fuck up a paedophile.'

In December 2013 it was reported that the UK's attorney general was due to publish guidance in response to a number of cases in which Twitter users had (apparently inadvertently) committed contempt of court by publishing inappropriate comments on (or information about) ongoing criminal cases. February that year, for example, saw the charging of people who had tweeted photos claiming to show child killer Jon Venables. That November the celebrity Peaches Geldof had apologised

for tweeting the names of the two women whose babies had been subject
to sexual abuse.

Peaches Geldof died at the age of 25 on 7 April 2014. Many of the
subsequent news reports observed that shortly before she died she had
posted online a childhood photograph of herself and her mother Paula
Yates, who had also died young as a consequence of heroin abuse. Bartlett
(2014) has pointed out that her death prompted 387 thousand tweets
within the first twenty-four hours of the news breaking. Bartlett suggested
that this process of 'dying online' might bear further scrutiny: 'how we as
a society incorporate social media into the process of dying and grieving
is an increasingly important subject.' Gold (2014) has added that 'this is
death in the digital age, where social media respond in seconds and old
media report for hours' – and that the online 'grunt of synthetic emotion'
in response to the death of this celebrity felt like 'entertainment posted
onto a random face.' To what extent then (we may be forgiven for asking)
might such an online response ever really be able to capture, for a frozen
moment, the *zeitgeist* of a culture, a nation or indeed the entire world?

Death by social media

Nelson Mandela died on 5 December 2013. He was considered by many to
the greatest person who had walked the face of the planet for at least the
previous two thousand years. His death did not go unnoticed in the media.
It was also an event which received a fair deal of attention from the users
of social media. Twitter and Facebook, in short, went wild.

Before examining the new media coverage of the event, it is perhaps first
worth looking at how more traditional journalistic institutions addressed the
story. Much of the conventional media coverage of Mandela's death might
itself be viewed as somewhat problematic. The most notorious instance
of such controversial coverage was a headline in the Italian newspaper *Il
Giornale* which described Mr Mandela as *il padre dell'apartheid* – the father
of apartheid. A BBC report on the evening of 5 December had meanwhile

noted that 'an international campaign was begun for the release of Nelson Mandela as around the world governments imposed sanctions on South Africa. In 1990 a courageous white leader F. W. de Klerk announced that the ANC would be unbanned.' The implication that it was the tenacity and courage of western nations and white South Africans which brought apartheid to its knees might, with hindsight, seem somewhat overstated. A similarly clumsy Sky reporter live at the scene of Mandela's Johannesburg home in the immediate wake of his death observed that 'everyone is wondering why people are singing and dancing, but that is how Africans react to death' and seemed impressed by the fact that a white man had spoken 'in fluent Zulu' – before turning to attempt to vox-pop a black South African mourner: 'Can you speak in English? No? Can you sing?'

The morning after Mandela's death *The Independent* devoted a page and a bit to a photograph of these South African mourners – an image in which black faces were strikingly absent. (By contrast, its sister paper – the abbreviated *i* – ran its parallel feature alongside an image of Mandela standing before a sea of predominantly black faces.) *The Times* was generally positive about Mandela's life but ran a couple of negative pieces about his legacy: 'Squabbling and petty jealousies blight the family name' and 'The ANC promised so much, but little has changed.' David Blair in *The Daily Telegraph* went further, arguing that 'as chief executive of South Africa's government, he was largely a failure. Mandela possessed neither the temperament nor the aptitude to run a modern government. Loyalty to the ANC and steadfast dedication to the anti-apartheid struggle won promotion. Genuine ability or even basic honesty seemed less important. Thus Mandela's cabinets were stuffed with time-servers and incompetents.' In its main feature on Mandela's life and death, the *Daily Mail* adopted an even more controversial approach. It emphasized his links with the more conservative forces of the British establishment: 'he admired the Queen – and Mrs Thatcher.' The paper also showed a picture of him meeting David Cameron in 2008 and stressed his friendship with the Spice Girls. It also spoke of his closeness to his white gaolers: 'when he left, he hugged these white men and told them how he would miss them.' Mr Mandela seemed indeed to epitomize traditional British values of aristocratic gentlemanliness: he was 'courteous' even to his enemies; he was a 'courtly, old-fashioned man who would tolerate most things except bad manners' – 'his family were royalty.' Yet the *Mail* also reminded its readers of his ex-wife Winnie Mandela's crimes and affairs; it recalled South Africa's culture of bribery; it noted that Mandela had been 'too old and worn out by 1994 to be an effective executive President' and depicted the 'preening members of the ANC' as a 'self-serving and permanent political elite.' In another piece (also on 6 December) the *Mail* columnist Max Hastings described Mandela as 'a weak executive ruler.' Eight days later, the paper noted the contribution of 'another great statesman, with her own claim to have played a pivotal role in South Africa's peaceful transition to majority rule' – Margaret Thatcher.

The *Daily Star* did little better, devoting half of its two pages on Mandela's death to a featurette on his association with the Spice Girls and David Beckham and another piece explaining how Prince William – on hearing of Mandela's death while attending the premiere of a biopic of the great man's life – had expressed his sadness at the news. *The Sun* went so far as to devote nearly half of its four pages covering the story to the news of Prince William's sadness. The *Express* meanwhile extended the better part of a couple of pages (again about half of its coverage of the event) to its account of the Prince's sorrow, but chose not to run the news of the former South African president's demise as its main front page story, opting instead to prioritize a report of inclement weather.

When one comes to examine the social media response to Mandela's death, one is struck by how the tone of much of the conventional news coverage may have misjudged the mood of a great part of the world online. Having said that, however, it should be acknowledged that the social media coverage of the event was not itself entirely without controversy. The *Huffington Post* reported on 6 December that celebrity heiress Paris Hilton had complained of a hoax tweet – purporting to come from her – which had confused Mandela with Martin Luther King: 'RIP Nelson Mandela. Your "I have a dream" speech was so inspiring. Amazing man.' The same day the *Spectator* and *Sun* columnist Rod Liddle caused what the *Daily Mail* described as 'widespread opprobrium on Twitter' when he wrote on his blog: 'For Christ's sake BBC, give it a bloody break for five minutes, will you? It's as if the poor bugger now has to bear your entire self-flagellating white post-colonial bien pensant guilt; look! Famous nice black man dies! Let's re-run the entire history of South Africa.' Four days later the *Daily Mirror* reported that people had again taken to Twitter to protest against the 'vicious, nasty attack' on the late South African President made by British National Party leader Nick Griffin – who had himself used Twitter to brand Mandela a 'murdering old terrorist.'

Twitter, of course, exploded with the news: *BBC News* alone posted 24 'breaking news' tweets on the subject within the first eight hours of the story breaking. Two days later the *Huffington Post* ran a story which collated many of the celebrity tweets on the subject. Nearly a quarter of those quoted (the likes of actor Samuel L. Jackson, businessman Richard Branson, singers Fergie and John Legend, and rapper Ludacris) alluded to having known Mandela personally; CNN anchor Anderson Cooper matched this name-dropping with his personal reminiscences of Soweto on election day in 1994. The American actor/musician (and apparently honorary Xhosa) Josh Groban referred to Mandela as 'Madiba' (his Xhosa clan name) and said that he would personally never forget the lessons he had learnt from him. The Barbadian singer Rihanna also appeared to adopt South African origins when she declared: 'you made your people proud!! We'll always love you for it!' The singer Queen Latifah meanwhile observed that Mandela's humanity had been unmatched specifically in *her own* lifetime (a curiously egocentric measure of his greatness). The musician R. Kelly somewhat idiosyncratically announced that 'God has chosen a soldier' – apparently unaware of the late saint's pedigree as an erstwhile peacemaker – while the musician's daughter Kelly Osbourne sounded an even more eccentric and hyperbolic note: 'Today is officially one of the saddest days. The world has lost the most inspirational person ever to walk the earth rip.' Sadly Twitter's 140-character limit had not allowed the daughter of the lead vocalist of Black Sabbath to elaborate upon: (1) the workings of the international organization (presumably under the auspices of the United Nations) which officially determines the sadness levels of days; (2) the comparative inspirational merits of Mandela, Buddha, Muhammad, Jesus, Marx, Lenin and Lennon; or, for that matter, (3) precisely why someone so extraordinarily inspirational had not inspired her to bother to capitalize the rather perfunctory 'RIP' she had tagged onto the end of her tweet. In a somewhat impersonal tweet the musician Paul Simon had spelt out

'Rest in Peace' in full, alongside the deceased's name and dates, but chose neither to offer his own perspective on Mandela's achievements nor to elaborate upon his own controversial breach of a United Nations cultural boycott of apartheid-era South Africa when he made his well-meaning and best-selling 1986 album *Graceland*.

For the most part, however, there was of course something emotional, personal and subjective at the heart of the common experience of the response to Nelson Mandela's death – it was not, for most people, about his history or his legacy, but was about the humanity of the man. Social media were therefore rather better positioned than traditional news organizations to foster and convey these messages; and Facebook was perhaps the best-placed of all social media to achieve this. (It may be noted that one of the least successful – because the most formal – websites in conveying the public reaction to the news was that of the Nelson Mandela Foundation itself: the day after it had announced the news of its namesake's death, it only displayed six comments posted in response.)

Alice Jones (2014) has suggested that that 'while Facebook is all about being inclusive and sharing – whether baby pictures or holiday memories – Twitter is harder, more exclusive and still mainly consists of specialists shouting at one another.' She has added that Twitter's '140-character limit demands wit and pithiness' but also 'breeds snark and trolls.' The profoundly personal public reaction to the death of Nelson Mandela was perhaps therefore best served by a site which had positioned itself as caring, sharing, communal and friendship-based, rather than one which emphasized such short-term wit and biting pith: one in which users for the most part purported only to address their presumed 'friends' (people they had chosen or acknowledged as their friends) rather than merely shouting to the world at large (or at least to those who have chosen – but who have not been chosen to be – their 'followers').

Because of the personal nature of this online reaction to the death of someone many felt they knew – because people did not appear to want geopolitical information or analysis so much as they wanted the opportunity to express themselves – it might be useful to give space to a personal account of the unfolding of this news from one particular Facebook user. (This Facebook user is, incidentally, me.) As the following is necessarily written from a personal perspective, it has been typographically differentiated in order to signal its distance from the purportedly academic style of the rest of this text:

Nelson Mandela died, for me, on Facebook. And indeed I had the strangest feeling on Facebook that Nelson Mandela had indeed lived and died for me, and for all of us who shared the experience through that platform. It was through Facebook, skimming Facebook that evening on my phone, that I discovered that he had died. Mandela was a virtual friend, an imagined friend, a mediated friend, to us all, and therefore Facebook became the perfect site to mark his loss. Facebook may not be capable of sustaining the long-term material and moral commitment of friendship, nor of establishing meaningful dialogues between opposing political or intellectual positions, but what the social media reaction to the death of Nelson Mandela demonstrated was that it offered a fitting arena for the mutual reinforcement of a broad liberal consensus, a temporary but powerful illusion of moral community, a place where – like the flower-strewn gates of Buckingham Palace in the immediate aftermath of the death of Princess Diana in 1997 – a shared experience of deeply and sincerely felt (albeit mediated, simulated or sentimental) grief could be further shared and intensified. What was important to the participants was the sharing of personal perspectives; but what seems significant to an observer was that most of these failed to result in extended dialogues (at most there would be comments relating to the original post, but it was rare that such comments related to and built upon other comments). Many users merely fell back on impersonal conventional formulae ('RIP, Nelson Mandela' and its variants), as if the fact that something had been said was actually more important than what it was that had been said – as if, in other words, they were registering their presence at this historic moment – as if daubing on the Facebook wall of history the words 'I was there.' Some chose simply to quote something that Mandela himself had said: I imagine a quick Google search or a glance through the early obituaries had been a quick way to glean an apposite aphorism. Others chose to express their personal perspectives on the life of the deceased – writing of how the great man had changed their own lives (he 'helped to shape me, politically' – he contributed to 'my real political awakening'), or of how they had come into contact with the nation he had shaped (in particular, reminiscences of trips to Robben Island). This then was not so much about Mandela as about themselves. It may be thought that there is something inappropriately narcissistic about this; or it might be supposed that this was the death that Facebook had been waiting for, the event which showed that Facebook users could appropriate anything at all as individually their own. Yet one Facebook friend of mine wrote eloquently of how Mandela had 'embodied and reflected our collective greatness' – and there was clearly a sense in these outpourings that this reflection of their own significance was something which many Facebook commentators were seeking to demonstrate. Another Facebook friend (in response to mass media and social media suggestions that we liberal westernizers had been primary collaborators in Mandela's struggle) commented that 'it was the struggle of the ANC not bourgeois boycotts that broke apartheid.' But, though we may have often been writing about ourselves rather than about Mandela, it was done in a shared spirit of love (albeit a form of love which inscribed self-love). One friend wrote, rather poetically: 'It seems somehow apt that the whole house is asleep, giving me a reflective moment.' That summed it up beautifully for me: that was also precisely what I felt. This was a time for quiet personal reflection, yet this was also a mode of reflection which I needed to share with the world. If there was something communally narcissistic in this mutual celebration of self-reflection, then this was only a bad thing if we think of it as something other than what it was. It was entirely real in – and only in – itself. And it seemed a great deal more sincere than what much of the mass media were doing at the time. Indeed even that night Facebook friends were already critiquing the mass media response: from the decision of Channel 4 News to give air-time to 'neanderthal Tories' who still considered Mandela a terrorist to the 'revisionism' of other politicians who had u-turned to claim Mandela as their own: 'history being rewritten live on TV' – 'tributes from self-serving political pygmies' – 'some of the people celebrating Mandela made the struggle against apartheid longer, harder and deadlier.' One friend was especially outraged by the BBC's claim that the song 'Nelson Mandela' had been written to celebrate his seventieth birthday. (He had only been 65 when it was released.) Another of my Facebook friends wrote insightfully about how social media platforms themselves had detrimentally influenced the broadcast media coverage of Mandela's death: 'crafted reflection about important lives appears to have been eschewed by reading out tweets from any old fucker.'

The Facebook response to the death of Nelson Mandela was not unequivocally unproblematic. But perhaps it suggests that there is some possibility that in the future such platforms might be deployed to capture and to bolster real-world forms of commitment – whether personal or political – or at least to foster informed public debate. Prior to Mandela's death – and much re-tweeted thereafter – a post on Twitter had suggested that the public beware of Tory hypocrisies in their responses to the impending death of Mr Mandela by trotting out a claim that in the mid-1980s David Cameron had been 'a top member of the Federation of Conservative Students, who produced "Hang Mandela" posters.' In response to this viral tweet, the website Skeptics Stack Exchange cited various sources (including the BBC and the *New Statesman*) which had suggested that Mr Cameron's high-level membership of such an organization would have been somewhat unlikely on the grounds that 'he wasn't politically active while he was a student and his political views were not extreme.'

If the worst of Web 2.0 is its propensity to aggravate hostility, abuse, misinformation and disinformation, then at its best it might prompt dialogues which challenge received notions of knowledge and power, and which thereby provoke directional shifts in civic conversations in favour of information-seeking, consensus-building and fundamentally radical solutions to societal and political problems, mobilizing a revolution in thought, word and real-world action.

CHAPTER 8

Revolutions

Natalie Fenton (Curran, Fenton and Freedman 2012) has interrogated the internet's claims to social and political empowerment through social networking sites and as an environment for the propagation of radical ideals. She has argued that, although 'social networking sites are heralded as [...] conferring agency' (2012: 124), it seems instead the case that 'social media work to reinforce already existing social hierarchies' (2012: 127). Furthermore, she has suggested that the illusion of agency offered by the internet may in fact dilute the possibilities of real-world empowerment: 'our experience of the internet itself may in some way actually hide what's going on [...] and blind us to the need for radical change' (2012: 141). She has exposed a number of clear problems with online political activism: that its convenient myths of empowerment may comfort us 'to the point of inaction' (2012: 170); that it is, for the most part, the province of a 'global middle class' (2012: 155); and that the political perspectives emerging from online activism do not for the most part develop constructive, consensual agendas, but tend to be limited to the emotive resistance to perceived practices of hegemonic power, and thus foster 'a politics of non-representation; a politics of affect and antagonism' (2012: 169). 'Why,' she pertinently asks (2012: 140), 'do we think the network of networks will somehow transcend previous inequalities, when the evidence on the ground is quite the opposite?' If we mistakenly think that this technology can save us, it may have a wholly opposite effect, reinforcing government and corporate power and fragmenting and diluting resistance to that power.

Springwatch

Despite Fenton's reservations, some popular credence has, however, been given to the idea that such sites as Facebook and Twitter facilitated and expedited the mobilization of pro-democracy demonstrators during the so-called Arab Spring, the wave of protests against authoritarian regimes which began in December 2010 – and that this (in the short term at least) represented a highly positive development. It remains unclear, however, to what extent this emblem of westernization – the social media platform – was seized upon by the western media as a way to make the story more accessible and interesting (and of course to facilitate media access to original source material, questionable though some of those sources might have been) – and to what extent these social networking technologies may be seen as promoting and enhancing civic discourse itself. The Libyan author Hisham Matar has, for example, argued that the role of social media in the Arab Spring has been overstated by media commentators (Singh 2011). Matar has suggested that 'the Egyptian uprising didn't happen on Facebook or Twitter because it couldn't have happened without the working classes, and they don't have access to those things. But it allowed the agile, internationalist elite to mobilise and play to the international media.' Matar has added: 'Social change takes a very long time. The internet is one of very many different tools and I don't think it's always going to make or break an uprising.'

Nevertheless, the series of related events which became known collectively as the Arab Spring began, for a short term at least, to foster some measure of a renaissance in cyber-utopianism. Such luminaries as Stuart Allan (2011) and John Downing (2012) reiterated an optimism for the democratic potential of new media in response to the uses of such technologies by democratic revolutionaries in this region for the mobilization of popular action. Khamis and Vaughn (2011), for example, also argued that 'the success of the Egyptian revolution, and the effective role that new media played in it, has broad implications [...] throughout the world.'

The news media in particular pushed this agenda. Baker (2011) wrote in *Time* magazine of how 'members of the new opposition [...] put up

Facebook pages and posted on Twitter [and] exhausted their thumbs sending text messages to everyone in their mobile phone books.' Williams (2011) pointed out in the *Daily Mail* that 'as in the uprisings that toppled long-time autocratic rulers in Egypt and Tunisia, Libyan activists are using social networking websites like Facebook and Twitter to rally protest.' Mark Almond (2011) penned a piece for *The Sun* newspaper unambiguously headlined 'How social networks trigger political downfall'.

Robert Fisk (2011) announced in *The Independent* that 'this is revolution by Twitter and revolution by Facebook.' Hugh Macleod (2011) in *The Guardian* reported on 'Syria's switched-on cyber activists.' Fawaz Gerges (2011) argued in *The Mirror* that the Arab 'revolt is powered by [...] cellphones, the internet, Facebook and Twitter.' *The Times*'s Stephen Dalton (2011) declared that Twitter was 'proving mightier than the sword throughout the Middle East' – while the same newspaper's Adam LeBor (2011) proposed that '140 characters can spark a revolution.' Indeed, on 13 February 2011 an editorial in *The Sunday Times* predicted that 'more revolutions will be fuelled by Twitter.'

Martin Fletcher (2011) in *The Times* reported from Egypt's Tahrir Square, where 'young men spray-painted Google, Twitter and Facebook logos on walls and tanks.' Fletcher referred to computer-technician-turned-revolutionary Wael Ghonim as 'the Facebook dreamer who led his nation.' Indeed Ghonim was later described by Ed O'Loughlin (2012) as a revolutionary who 'felt the need to tear himself away from the action because he wanted to update his Facebook page.' Though Ghonim (2012) would later emphasize the role of new media in his account of what he calls *Revolution 2.0*, his words that day in Tahrir Square were perhaps more illuminating: 'I liked to call this the Facebook Revolution, but after seeing the people out there I think it's the Egyptian people's revolution' (Fletcher 2011).

In his own memoir of the Egyptian uprising Ghonim referred repeatedly to the shortcomings of new media in terms of their revolutionary impacts. He noted for example that members of online campaigning groups tended to be young while older Egyptians remained away (Ghonim 2012: 113). He pointed out the problems faced by the online revolutionaries when Facebook chose to suspend certain campaigning pages because of copyright violations or the use of fake identities essential to the campaigners' safety (Ghonim 2012: 113–119). He emphasized that, while the posting of personalized content on social networking sites could inspire people in ways that the facts and statistics deployed by human rights campaigners could not, he was not suggesting that the former could ever replace the latter (Ghonim 2012: 88). He also wrote of the need to use these sites not only for the purposes of 'communication and coordination' but also to 'promote a culture of dialogue [...] and to cultivate a tradition of tolerance' (Ghonim 2012: 155, 113). Ghonim's problematization of these media forms did not, however, entirely diminish the premature cyber-enthusiasm of some contemporary western commentators.

The western media's attempts to appropriate these revolutions as offshoots of western media technologies and cultures may to some extent be explained by their desire to atone for (or to cast a veil over) the apparent indifference of many western media organizations (and many western governments) to the decades of human rights abuses perpetrated by these Arab states: from Tony Blair's visits to Colonel Gaddafi to the BBC's screening of a documentary funded by the Mubarak regime. There seems something meanly self-aggrandizing in the West's claims to the liberatory potentials of its own technologies. For, as Doreen Khoury (2011; 84) has supposed, the 'ownership of the Arab revolutions will always belong to the Arab people and not to Facebook or Twitter or any of the other online tools.'

As Chebib and Sohail (2011: 155) have suggested, social media cannot be seen as 'a trigger for the revolutions.' They have stressed that in the Egyptian uprising of 2011 'social media's main role was as a facilitator and an accelerating agent.' Courtney Radsch (2011: 80–81) has pointed out that in the months leading up to Egypt's uprising the Egyptian blogosphere, while reflecting a political situation that was clearly combustible, lacked a revolutionary spark – and that this spark was provided by the Tunisian uprising – despite that fact that such combustibility was not particularly evidenced in Tunisia's own blogosphere. We may infer, therefore, that it was not blogosphere itself which set the region alight. Indeed Morozov (2011b) has added that while 'it's been extremely entertaining to watch cyber-utopians [...] trip over one another in an effort to put another nail in the coffin of cyber-realism' those cyber-utopians who think the Arab

Spring was ignited by activities on social networking websites are ignoring 'the real-world activism underpinning them.'

In September 2011 the BBC's Mishal Hussain presented a two-part documentary about the Arab Spring entitled *How Facebook Changed the World*. What was notable about Hussain's documentary was how (despite its title) it demonstrated that social networking sites were perhaps most significant not in fomenting revolution internally but in their capacity 'to show the outside world what was happening.' Hussain also repeatedly emphasized the limitations of the idea of the virtual revolution. She pointed out that in Egypt only 20 per cent of the population had internet access, and explained how messages were sent not through the electronic ether but via taxi drivers: when the Egyptian authorities blocked the internet, 'the activists already had their plan and technology was no part of it.' She added that when the Libyan government made a similar move, Libyan rebels also reverted to 'old technology' – driving to the border with their video footage to get their messages out. Hussain's documentary concluded with a broader question: the question as to whether the same technology that helped these nations break with the past could 'be harnessed for a better future.'

Ghannam (2011: 23) has suggested that 'blogging and social networking alone cannot be expected to bring about immediate political change' and that we should therefore focus not upon the headline-grabbing online drama but upon 'the long-term impact, the development of new political and civil society engagement, and individual and institutional competencies.' James Curran (2012: 52) has added that though it has been claimed that social media 'enabled flash demonstrations to take place, and encouraged protests to spread across national frontiers [...] this analysis foregrounds the drama of the uprisings and the enabling role of communications technology, while paying little attention to the past or to the wider context of society.' Curran has then concluded that 'the Arab uprisings were the product not of Twitter but of dissent fermented over decades.'

Sorice and De Blasio (2014) have pointed out that 'we cannot say that social media alone can make a revolution.' Magetts and John (2014) have argued that although 'we know that social media played a role' in the mobilization of the Arab Spring it is now 'difficult to see exactly what that role was' – beyond the fact that their influences thereon were 'unruly and unpredictable.' They have argued that the distribution of such effects follow patterns similar to earthquakes: they form 'a chaotic system like the weather.' They have judged the resultant mobilizations of political activity to be unstable and of questionable sustainability.

Prentoulis (2014) has pointed out that the use of social media by protest movements constructs an 'antagonistic political frontier' between the people and the establishment. Internal fragmentation on the side of the protest movement itself further undermines opportunities for the development of a 'collective discourse' which might 'become institutionalized and bring change about.' Indeed that very process of real-world concretization – when an informal protest movement attempts to solidify into a formal political party informed by a positive agenda of constructive action – tends to generate unintended tensions between the horizontal communication structures of social media usage and the vertical communication hierarchies deployed for party management. Whether in Prentoulis's own example of the economic protest movements in austerity-hit Greece, or in the case of the Egyptian revolution and its aftermath, or even in relation to the discontinuity between Barack Obama's uses of social media platforms as fora for agenda-setting debates in the processes of election campaigning and in the contexts of his actual administrations, these tensions define the failings, thus far, of these forms to fulfil their much-vaunted promise and their actual potential to contribute to a renaissance in participatory democracy. There are clearly opportunities for social media to be deployed in more effective ways to inform the consensual development of political agendas and policies, but this process would require: (1) full and equal popular access to, and uses of, these platforms, and to the sources of accurate information necessary to support public debate; (2) a culture shift in the structures and uses of these platforms to promote constructive dialogue and engagement with the full societal diversity of voices; and (3) an acceptance by the political classes (including revolutionary leaderships) that such dialogues should directly inform political debates – rather than merely that political leaders should give the public the impression that they are listening. This utopian situation might transform a mode of representative democracy from which the electorate appear increasingly alienated into a directly participatory democracy; but any such shift would clearly be a long way off, and would be dependent upon massive and evolutionary political and societal changes, rather than merely upon an illusory quick fix offered by the miraculously convenient existence of this medium of communication.

The experiences of the Arab Spring do not, in retrospect, suggest that social media offer the key in themselves to progressive political change. It has become increasingly clear in the Arab region's post-revolutionary nations that Web 2.0 has not established a dialogical political consensus, and that the public sphere, such as it is, remains a violent, turbulent and resoundingly material space. While the lengthy period of bitter civil war experienced in Syria, and the rise therein of militant Islamic extremism, have done little to revive hopes for the long-term successes of these social media revolutions, events in Egypt have, in particular, subdued the cyberoptimists' earlier enthusiasm. In July 2013 the Egyptian military ousted the nation's democratically elected president Mohamed Morsi (a post-revolutionary president whose Islamist credentials did not necessarily impress all of the middle class revolutionaries who had inadvertently swept him to power), and in January 2014 introduced a new constitution. Yet oddly this counter-revolution did not appear entirely antagonistic to the logic of the Facebook revolution and its advocates, but for some (for a while) seemed to represent a consolidation of the strategies of the paradoxically elitist populism of that movement. On 4 July 2013 the BBC observed that 'large numbers of Egyptians who [once] saw the military as a roadblock to democracy are now casting the army as its champion.' It added that 'a notice on Mr Morsi's Facebook page denounced the army for its military coup.' A reminder then that those brought to power by social media may have their falls from grace memorialized the same way – just as those (like Peaches Geldof) who live by social media may die by social media too.

In April 2014 *The Economist* reported on the imprisonment of three Egyptian political activists whom it described as having been among 'the best-known youthful faces of the 2011 uprising.' The three had been peacefully protesting a law which banned the gathering of ten or more people without police permission. This 'stifling of opposition' had come to seem typical of the new regime. The demagogic and monologic dictatorship of social media had eventually fostered the political logic of the nation's reversion to totalitarian rule.

Within this context, it seems that new media technologies may not have offered an impetus for democratic development so much as a catalyst for the mobilization of flash demonstrations and for the dissemination of information and images to international news organizations. Social media had, in short, failed to deliver on the one thing which they had promised – the one thing which made them unique: the potential to support civic interactivity and to promote dialogue and the development of social and political consciousness and consensus.

Summer of discontent

At around the same time as the online emergence of revolutionary dissent in the Arab region, a number of anti-capitalism protests had begun in the UK, particularly in the area of St Paul's Cathedral in the City of London. These too may be noted for their instability, unpredictability and unsustainability. Similar protests were seen on Wall Street in New York. Many commentators questioned the point of these protests – not their sincerity, but the fact that they did not have a defined goal. Like the protests against economic reforms that had regularly engulfed Athens for the previous few years, they appeared more like an outpouring of anger than a focused attempt to drive a particular political or economic agenda. These clashes and demonstrations seemed to demonstrate that while new media technologies were clearly useful as communication tools in mobilizing protesters to action, their usefulness in establishing strands of dialogue which might coalesce into specific strategies – in other words, their promotion of a public sphere of socio-political debate – appeared limited.

Also at this time, much was made by the British mass media of the deployment of social media platforms in the fomentation of other forms of civil unrest – a series of riots which engulfed English towns and cities in August 2011. Jackson (2011) has suggested that 'as riots spread across

England [...] so too did a wealth of misinformation about them, fuelled by social networking sites like Twitter [...] As the rioting moved to other parts of London and England, inaccurate – at times inflammatory – information began appearing on social media sites [...] In times of uncertainty or heightened fears in communities, rumour mills have always churned. But in a world of social media, a whisper can acquire a damaging momentum regardless of its relationship to the truth.' Several individuals were gaoled for their use of social media platforms to incite these riots: one, for example, had launched an event on Facebook entitled 'Smash down Northwich town' while another had started a Facebook page called 'Let's have a riot in Latchford'. Yet another was given a four-month custodial sentence after suggesting 'Let's start Bangor riots' on Facebook.

These riots began in London on 6 August 2011 and then spread across other British cities (including Liverpool, Manchester, Birmingham and Bristol) over the following four days. During this short period the *Daily Mail* (out of its 67 stories about the riots) featured 16 stories which held new media in some way accountable for this outbreak of urban violence – as opposed to only 14 stories which blamed left-wing and liberal politics (a traditional target for the *Mail*) and just three stories which portrayed the BBC – another perennial *Mail* target – as at fault. On 8 August 2011 the *Mail* ran a story headlined 'Tweeters fan the flames of hatred.' The following day the *Mail* announced that 'Young looters coordinate raids via Twitter and BlackBerry.' The next day another *Mail* headline denounced the 'BlackBerry ringleaders.' The day after that, a feature article by Tanith Carey asked 'Why do people become so vile online?' Carey conjectured in this article that 'the downside of the free speech offered by blogs, Twitter and social networking is that it has created a generation of narcissists obsessed with their own opinions.'

Drawing a direct comparison between UK rioters and Islamic terrorists, the *Mail* pointed out on 9 August that 'several countries [...] have complained that BlackBerry messages [...] may be used by terrorists.' Just as the rioters' appropriation of these technologies might be equated with acts of terrorism, so they might be addressed by online surveillance techniques once (in the days before Edward Snowden's revelations) more often associated with counter-terrorism strategies. As the *Daily Mail* announced on 9 August, 'intelligence experts are scouring public messages for evidence of those behind the mayhem' – and on the following day: 'government spies have been drafted in to track riot ringleaders who have been using encrypted instant messages on their BlackBerry smartphones to avoid detection.'

The Sun newspaper's coverage of the riots over this five-day period vented a similar sense of outrage in relation to these uses of social media. Only two of the 58 stories which its national edition ran on the subject blamed the political left – while 11 portrayed new media negatively in relation to the riots. Its Scottish edition was particularly forthright in its condemnation of the social media contribution to the chaos: 'Thieves used Twitter to urge looting' (8 August), 'Nail the Twitter rioters' (9 August) and 'Scots pair held over Facebook riot pages' (10 August).

The emphasis on the role of interpersonal media technologies in these events was somewhat unprecedented. Commentaries on the Brixton riots of the 1980s had not, for example, focused on the telephone as the cause of the unrest, nor had the early postal system been commonly held to blame for Peterloo. In part it may be supposed that certain elements in the British press chose to blame the uses of new media technologies for this civil unrest in preference to addressing such issues as the growing sense of socio-economic injustice among a disaffected and excluded youth as possible root causes of these events (a phenomenon, for example, identified by *Reading the Riots*, a study conducted by the London School of Economics in collaboration with *The Guardian* newspaper). In their ensuing hysteria the popular press also seemed entirely unaware of their own role in the spread of panic and chaos across the nation.

It may also be supposed that the UK's news institutions' ambivalent but emotive perspective on new media technologies may to some extent explain their praise of the influence of interactive technologies in the Arab rebellions and their almost simultaneous condemnation of that influence in the British riots. Journalistic organizations, while obliged to get into bed with new media technologies (in their exploitation of online publication, for example, and of user-generated content), appear to continue to feel professionally, institutionally and economically threatened by the rise of this prospective Fifth Estate. While Jones and Salter (2012: 173) – and many others – may suggest that the future of journalism lies in its integration of online technologies, a certain degree of scepticism and paranoia clearly remains.

A similar ambivalence appears to apply to academic perspectives on new media technologies: perspectives which are at once optimistic about their potential to promote a democratic and egalitarian future envisaged by the liberal ideals dominant in so much of humanities and social science academia, and to enhance the dissemination of the fruits of academic endeavour (by sponsoring a far broader reach of education and publication than previously possible), while at the same time cautious as to their capacity to de-institutionalize or de-professionalize intellectual authority. Just as the personal news-blog may threaten to undermine perceptions of professional journalistic authority, so the likes of Wikipedia might call into question the public authority of the university academic or of the peer-reviewed journal.

Indeed it may further be noted that one man's freedom fighter is another man's rioter; and that therefore the perspectives of the media establishment upon civil unrest may in the end come down to how close (geographically and economically) that unrest comes to undermining their own established interests: that violence in Egypt, for example, represents to the UK media establishment (and, for that matter, to the UK academic establishment) legitimate political protest (and therefore the technologies which appeared to facilitate this violence were seen as liberating), while violence in London represents wanton criminality (and therefore those same technologies were in this case seen as morally corrupting and corrupt).

It might even be suggested that, in the case of the 2011 riots, what has really incensed certain elements of British journalism is the appropriation of certain new media technologies (once considered the province of a professional elite) not only by the general population but specifically by groups of British citizens perceived as representing an underclass – those so-called *chavs*. Just as this young precariat have appropriated Burberry (once a symbol of moneyed power) so they have now usurped BlackBerry (a similar symbol): and, as the British press was wont to emphasize, they specifically *stole* these phones during their rioting sprees. (In a criminological chicken-and-egg paradox, it remains unclear which came first: the looting of the phone or the use of the phone to organize the looting.)

But if the appropriation of these communications technologies really so threatens the institutions of established power that this appropriation has become a determinant in the outraged denigration of a disempowered section of society, might then these technologies in fact eventually offer to empower that underclass and overthrow those entrenched structures of power?

Clicktivism

On 15 November 2012 *The Independent* supposed that 'social media like Twitter and Facebook are a new front in the war for Gaza.' The paper pointed out that after 'the Israeli Defence Force parked its tanks on the social media site boasting at @IDFSpokesperson that Ahmed al-Jabari, head of the Hamas military wing, had been eliminated by an Israeli air strike' the armed wing of Hamas, Izz ad-Din al-Qassam Brigades, tweeted back to proclaim: 'Our blessed hands will reach your leaders and soldiers wherever they are (You Opened Hell Gates on Yourselves).' The following year, in September 2013, terrorists in Nairobi provided (in the words of the *Daily Mail*) 'a live commentary on Twitter as they murdered dozens of innocents at Westgate shopping centre.' It is clear that even those willing to deploy the most violent means in their attempts to effect their visions of political change also appear to value social media as increasingly useful tools to supplement their armed struggles. (Cf. Varandani 2014; Vincent 2014a.)

Despite the uncertainties surrounding their efficacy and their longer-term effects, the popular notion that social media might in themselves foster revolutionary societal developments has had far-reaching geopolitical influences. In April 2014 it was, for example, revealed that between 2009 and 2012 the United States Agency for International Development had run a text-message-based social networking system described as a 'Cuban Twitter' in a bid to stimulate on that Caribbean island a revolutionary fervour intended to emulate that of the Arab Spring.

Even if there is nothing innately revolutionary about the elitist-populist power of social media, many governments (from the United States to Turkey) may well believe (either hope or fear) that there is. In March 2013 it was reported that Turkish Prime Minister Recep Tayyip Erdogan had vowed to 'wipe out' Twitter after Turkish citizens had used the site to spread allegations of corruption in his government. A few days later a Turkish court lifted the ban on Twitter; but the following day the Prime Minister banned YouTube instead, after what purported to be a leaked recording of Turkish government officials discussing the Syrian crisis appeared on the site. Shortly thereafter, at the start of April, another court rescinded the YouTube ban. The social media revolution is, it seems, inevitable: the more that you attempt to regulate or prohibit the uses of such forms, the more they proliferate. As the BBC reported on 21 March 2014: 'Turkey is one of the world's most active countries on social media. So what happens when Twitter is blocked there? Millions of tweets from Turkey, that's what.' But what, one might ask, is the impact of so many tweets?

The question is not so much whether the social media revolution can be prevented or curtailed: the question is how truly revolutionary it will be. Will online activism (and internet-mobilized activism) propagate progressive or reactionary trends? These technologies are, of course, ideologically neutral in themselves; but if we see them as inherently progressive (if we see their uses as progressive in themselves; if we consider our actions progressive merely because we are employing these platforms) then their impact may well be reactionary – that is, their impact may well end up reinforcing entrenched structures and practices of power.

Easton (2012) has supposed that 'Facebook and Twitter campaigns, it seems, are replacing the old-fashioned demo or sit-in.' The cyberoptimist may point to one of those rare but celebrated occasions upon which an ordinary individual has won a minor victory against a multinational corporation or public sector organization (usually a refund or an apology), thanks apparently to an online campaign or petition. The cyberpessimist might counter this example with the arguments that: (1) the rareness of such examples demonstrates that they are exceptions which highlight the general rule that large institutions are overwhelmingly more powerful than individuals; (2) it is in the interests of such institutions that such exceptions are so publicly celebrated in order to invest their consumers with a sense of active empowerment which will continue to underpin their passive consumer behaviours; (3) this ongoing confusion of consumer behaviours with citizenship rights and responsibilities undermines the possibilities of democracy; (4) we do not all have the singing, song-writing or filmmaking skills to make our online campaigns appeal to broad sections of the clicking public; (5) if we all did, then it would take even more to make the average surfer stop and click (we cannot *all* be remarkable); and (6) such clicktivism has remarkably little influence precisely because it demands remarkably little commitment: a million people clicking to support an online petition are unlikely to change their real-world buying or voting patterns, while a few thousand people marching through a city centre on a rainy Saturday afternoon may well demonstrate sufficient commitment to their cause to indicate other changes to their behavioural patterns as consumers or citizens.

Dennis (2014) has, however, offered a compelling critique of some of the failings of simplistic denunciations of that lazy clicktivism dubbed 'slacktivism' – that 'low threshold of political action online such as signing an e-petition, clicking *like* on a Facebook page or sharing a video' – as themselves being symptomatic of an intellectually unrigorous and dismissive cyber-dystopianism. One might nevertheless recall that a convincing deconstruction of a particular argument for atheism is not on its own sufficient to prove the existence of God. Dennis's assertion that online political engagement may offer 'important democratic short cuts' appears to take as read the notion that shortcuts towards democracy will necessarily lead to the same kind of democracy as the long-trodden road. But if democracy is a direction of travel or a societal aspiration – rather than an end ever achieved or ever achievable – if democracy then represents an ongoing and dynamic range of possibilities, we may see the journey (or the open-endedness of the process) as the end in itself, and, if so, then shortcuts may be seen as less sustainable or beneficial than attempts simply to make the journey and the conversation last. Short cuts to democracy might in those terms turn out to be diversions towards a dead end.

But might it be argued that Web 2.0's emphasis on user-generated content represents a radical (if eventually narcissistic and solipsistic) individualism which undermines the stability of hierarchical structures *per se* (shifting order towards chaos: dictatorship towards democracy, democracy towards anarchy)? That might go some way to explaining why the same platforms might lead both towards revolutionary activity and to riots. What the internet promised us, in our cyberutopian dreams, was the possibility of a new and liberating consensus based upon dialogue and mutual understanding; but what it has delivered – from London riots to the Arab Spring, from anti-capitalist protests to cyberbullying – is a contentless violence born out of the frustration of anger without an agenda (without a realisable and workable agenda; that is, without a shared vision). No new global enlightenment has as yet been bred from a digital public sphere; rather, the diverse and directionless voices of these narrow-cast private spheres have been unable to heed each other in the chaos of their irresponsive monologues. The promise that new media technologies might sponsor a liberation from totalitarian or authoritarian regimes is true only insofar as liberalization is the first stage of anarchization; and if this clamour of unlistening voices might deconstruct modes of oppressive authority and move oppressed societies towards forms of liberal democracy, then it also moves liberal democracy towards the anarchy of unstructured and anonymous (and therefore unaccountable) power. Or rather it only does so until structures of power (either traditional or new) reassert themselves – in omniscient hierarchies which are more anonymous, less visible and therefore even less accountable (although possibly more demagogic) than ever before, and which by claiming to offer empowerment in fact diminish popular resistance to their power.

The deployment of digital media technologies towards a project which calls itself democratizing is not, then, necessarily conducive to the progress of democracy. *Democracy* is a term which is often invoked in defence of its opposite: after all, how many totalitarian dictatorships have over the last half century or so described themselves as democratic? Wikipedia co-founder Larry Sanger has commented that there is clearly a problematic ambiguity in the ways in which we speak of the processes of democratization: 'it depends on what you mean by democratization. Does the question concern progress toward progressive political ideals, for example, or toward basic and literal democracy (which isn't much of a concern in our countries), or toward a greater degree of empowerment of the formerly silent individual in the marketplace of ideas – or what?' This nebulous term is repeatedly used to legitimize a variety of very different positions. Is it, for example, democratic that new media give their users a voice, even though it may be that nobody listens? Is it democratic to promote self-expression without any attempt to evaluate (and therefore to value) what is expressed? One might suppose that what one would like to mean by democratization represents the civic empowerment of the individual – and one might therein argue that we are empowered not only by access to the tools of self-expression, but also by access to sufficient knowledge to allow us to form and express our judgments through critical and contextualized reflections which might then promote informed debate. In other words, what we may mean by 'democratization' may bring us back towards the promotion of what Jürgen Habermas would call a public sphere – an open, dynamic and transparent forum for public dialogue.

The Silicon Valley entrepreneur turned internet apostate Andrew Keen has, however, suggested that Web 2.0 has little to do with democratization. He has said that, when cyberenthusiasts speak of the internet's powers of democratization, they appear to be suggesting that 'it flattens the playing field – everyone has the same voice – it breaks down the hierarchies, the supposed oligarchies of the old media world.' Yet Keen has supposed that 'words like democracy get thrown around all the time, but not everyone agrees about the ideal of democracy.' He has postulated that the internet's magnates have themselves formed 'the internet's digital elite' and has criticised this elite's hypocrisy: 'one of the annoying things is that they continue to speak in the voice of the people, but they are the elite. The counter-culture has become the establishment. The new elite has seized power in the name of the people.' Keen has stressed that this situation represents 'a new challenge to representative democracy.'

At the end of George Orwell's novel *Animal Farm* the leaders of the revolution – the pigs – join forces with the human elite. Orwell (1951: 120) famously wrote that the other animals 'looked from pig to man, and from man to pig, and from pig to man again; but already it was impossible to say which was which.' The conclusion to Orwell's fable offers two messages which may still seem pertinent to the internet generation. The first is obvious enough: that, as power corrupts, the revolutionary counter-culture itself develops an elitist hierarchy.

But we can perhaps see a second level of meaning in Orwell's image: the tragedy is not only that pig and man are objectively indistinguishable; the tragedy is that their audience can no longer distinguish between them. And so we view the new elite as our liberators and see liberation in the new hierarchies; we mistake our unheard vocality for power; we accept the rumours, Chinese whispers and more malicious fabrications which arrive through our machines in lieu of the tested, grounded, validated knowledge which would offer the only actual possibility of our empowerment.

The pursuit of happiness

That truly remarkable book, that imaginary prototype of new media interactivity, that fantastical forerunner of the populist electronic encyclopaedia, Douglas Adams's *Hitchhiker's Guide to the Galaxy* famously noted in the late 1970s that the planet Earth was *mostly harmless*. We might say the same thing about new media technologies. They only become harmful when we convince ourselves that they are, in themselves, doing us good. It is that assumption which may be preventing us from properly understanding their impact and therefore from realizing their potential to improve our societies. The central role which these technologies play in our lives might thus eventually facilitate our social and political interactions in ways aligned to a consensual notion of progress towards the enhancement of the public good, and might propel media and policy agendas beyond the reactionary constraints and hierarchies of entrenched institutions steeped in elitist and exclusionary structures of power.

Yet, insofar as it has promised swift, easy and ultimately unsustainable solutions to our social, political and existential woes, the new media revolution has so far provided a comforting but fundamentally counterproductive distraction from these issues rather than the practical facilitation of the development of progressive agendas designed to address them.

Perhaps, then, we need to stop looking for easy answers and start facing up to complex questions and difficult choices. One way to start this process, this book has suggested, it to acknowledge the incoherence and absurdity of our current situation.

The nineteenth century Danish philosopher Søren Kierkegaard argued that we can only counter despair through a recognition of absurdity. Kierkegaard recalled the story of Abraham and Isaac, the Biblical account of the father instructed by God to sacrifice his son, the man who met that divine paradox with the paradoxical absurdity of his own faith. Kierkegaard (2006: 49) observed that the triumphant absurdity of Abraham's faith lay in the fact that he as a single individual became 'higher than the universal' – an absurdity mirrored for Kierkegaard in history's greatest absurdity, the incarnation of divinity as an ordinary individual human being (Kierkegaard 2009: 117).

Roland Barthes (1977: 134) witnessed something of similarly transcendental significance when he observed how weakness defeated strength in the Biblical story of Abraham's grandson Jacob's struggle with the angel. The utopian myth of new media interactivity has offered us something similarly – almost impossibly – absurd: the ordinary individual made equivalent in power and significance to the totality of society; the mobilization of the individual within effective processes of democratic participation; the dialogical development of a socio-political agenda by consensus; the apotheosis of all ordinary individual human beings as avatars of social, cultural and political agency. This myth might be said to parallel another famous Biblical account of weakness beating strength, the story of David and Goliath – if, that is, instead of killing Goliath, David had invited him to sit down and develop a dynamic, interactive, pluralist and egalitarian process for the establishment of consensual decision-making, one in which all voices, big and small, carried equal and essential weight. This, in effect, is what evangelical cyber-utopianism has promised us; and yet what has been delivered by so many of the political and commercial interests which have come to dominate the new media revolution has been the precise opposite of this, a mere sop to any popular desire for empowerment.

The myths of Abraham, Jacob and David represent the individual's arduous, risk-laden and almost impossible struggle against massive power as being both absurd and, at the same time, definitively human. To acknowledge our disempowerment (indeed, to acknowledge the absurdity of the possibility of our empowerment) may eventually afford our only possibility of empowerment. But could we imagine these possibilities without the specious buttress of a transcendental structure (be that of God 1.0 or of Web 2.0)? To embrace again what Albert Camus deemed the splendour of this absurd world may allow us to recognize the absurdity of our desires for empowerment, and therefore permit us to differentiate between, on the one hand, the agonisingly slow, incremental steps which we may take (the overwhelmingly unrewarding efforts of actual participation we must make) towards the slim possibility of their eventual fulfilment, and, on the other, the spurious yet seductive claims advanced by commercial, political and religious interests as to easy routes through which (they declare) we are already being empowered. In essence, our only possibility of empowerment lies in our ability to distinguish between these two absurdities – the strenuous absurdity of individual empowerment as experienced, in fable at least, by Abraham, David and Jacob, and the absurd illusion of mediated empowerment as offered by Cowell, Wales and Zuckerberg – and then to embrace the former, comfortless state. This is not a passive acceptance of powerlessness but an active realization that material empowerment requires effortful commitment.

Having returned to Albert Camus, it is perhaps appropriate to conclude by recalling Camus's interpretation of the tale of a dead Greek king which concludes his great essay on absurdity, *The Myth of Sisyphus*. Camus finds, in the story of the man condemned for all eternity to roll a boulder to the top of a mountain, watch it roll back down again and then roll it up again, an apt enough analogy for the absurdity of the human condition. Yet Camus (1975: 111) discovers the possibility of nobility and joy in this Sisyphean situation: 'The struggle itself towards the heights is enough to fill a man's heart. We must imagine Sisyphus happy.'

The Sisyphean futility of the digital game's disempowering escapism or of reality television's fifteen minutes of fame, or for that matter of the dilution of culture, society and democracy witnessed in the projects of Wikipedia, Facebook, Twitter and electronic politics, may most usefully be countered, from Camus's perspective, by a recognition of that futility.

This recognition may, eventually, be the extent of our potential for liberation. Or it may, after all, be only the beginning.

The acknowledgement that a certain degree of unhappiness may represent a perfectly sincere, rational, healthy and highly likely response to any particular set of social conditions (or indeed to the human condition *per se*) may offer an essential first step towards the recognition that it is not the only such response imaginable. It is only by accepting the near impossibility of happiness that such happiness may become, albeit improbably, possible. For, as Kierkegaard (2008: 27) suggests, 'he who says without pretence that he despairs is […] a little nearer […] being cured than all those […] who do not regard themselves as being in despair.'

For Camus the absurdity that we must embrace is that of a world without innate purpose or meaning; for Kierkegaard the absurdity that will redeem us is the leap of faith which projects us beyond the resulting despair. A recognition of the inevitability of absurdity does not in itself determine a pessimistic or optimistic perspective, a dystopian or utopian path. It merely offers us the possibility of that choice. That is the extent of the liberation afforded by the acknowledgement of the absurd; it is the extent of any possibility of freedom at all. Technologies cannot save us; it is what we choose to do with them that matters. What matters is whether or not we choose to do the difficult things – to make the effort.

It is through such choices that society and politics may be transfigured into ideational mirrors of one another (politics becoming the ideal of society; society becoming a model for politics), and aspirations towards interactivity may, in fostering dialogical trust between diverse publics and structures of power, be transformed into the dynamics of democracy – a historical movement whose direction of travel is perhaps underpinned by something which ancient philosophers once liked to call friendship or love, but which we might prefer simply to see as in essence the most social medium of them all.

Bibliography

Adams, D. (1979). *The Hitchhiker's Guide to the Galaxy*. New York: Harmony Books.

Adams, E. (2009). 'The philosophical roots of computer game design' in *Under the Mask: Perspectives on the Gamer*, University of Bedfordshire, 5 June 2009.

Adorno, T., and Horkheimer, M. (1979). *The Dialectic of Enlightenment* (trans. Cumming, J.). London: Verso.

Alas, J. (2007a). 'Card readers the only challenge in e-election' in *The Baltic Times*, 7 March 2007.

Alas, J. (2007b). 'Thumbs up for mobile voting' in *The Baltic Times*, 3 October 2007.

Allan, S. (2011). Keynote address in *Political Studies Association Media and Politics Conference*, Bournemouth University, 3–4 November 2011.

Almond, M. (2011). 'How social networks trigger political downfall' in *The Sun*, 27 January 2011.

Althusser, L. (2006). *Lenin and Philosophy* (trans. Brewster, B.). Delhi: Aakar Books.

Anderson, J. (2011). *Wikipedia: The Company and its Founders*. Minnesota: ABDO Publishing.

Apter, T. (2010). 'Internet pornography' in *The Observer Magazine*, 10 April 2010.

Aristotle (2004). *The Nicomachean Ethics* (trans. Thomson, J.). London: Penguin.

Arsenault, D., and Perron B. (2009). 'In the frame of the magic cycle: the circle(s) of gameplay' in *The Video Game Theory Reader 2* (ed. Perron, B., and Wolf, M.). London: Routledge. 109–131.

Ashton, J. (2012). 'Silicon supremo' in *The Independent*, 19 May 2012.

Åström, J. (2004). 'Digital democracy' in *Electronic Democracy* (ed. Gibson, R., Römmele, A., and Ward, S.). Abingdon: Routledge. 96–115.

Auden, W. H. (1979). *Selected Poems*. London: Faber & Faber.

Ayers, P., Matthews, C., and Yates, B. (2008). *How Wikipedia Works*. San Francisco: No Starch Press.

Baker, A. (2011). 'How Egypt's opposition got a more youthful mojo' in *Time*, 1 February 2011.

Baloun, K. (2006). *Inside Facebook: Life, Work and Visions of Greatness*. Self-published.

Bang, H. (2010). 'A new ruler meeting a new citizen: culture governance and everyday making' in *Governance as Social and Political Communication* (ed. Bang, H.). Manchester: Manchester University Press. 241–266.

Barthes, R. (1974). *S/Z* (trans. Miller, R.). New York: Farrah.

Barthes, R. (1977). *Image-Music-Text* (trans. Heath, S.). London: Fontana.

Bartlett, J. (2014). 'Peaches Geldof, Twitter grief and the strange, poignant phenomenon of dying online' in *The Daily Telegraph*, 9 April 2014.

Bates, L. (2013). 'Does Facebook have a problem with women?' in *The Guardian*, 19 February 2013.

Baudrillard, J. (1988). *America* (trans. Turner, C.). London: Verso.

Baudrillard, J. (1994). *Simulacra and Simulation* (trans. Glaser, S.). Michigan: University of Michigan Press.

Baudrillard, J. (1995). *The Gulf War Did Not Take Place* (trans. Patton, P.). Sydney: Power Publications.

Baudrillard, J. (2005). *The Intelligence of Evil or The Lucidity Pact* (trans. Turner, C.). Oxford: Berg.

Bell, E. (2014). 'Info wars: time for Google & Co to come clean' in *The Guardian*, 26 May 2014.

Benjamin, W. (1992). *Illuminations* (trans. Zohn, H.). London: Fontana Press.

Bignell, J. (2005). *Big Brother: Reality TV in the Twenty-First Century*. Basingstoke: Palgrave Macmillan.

Biressi, A., and Nunn, H. (2005). *Reality TV: Realism and Revelation*. London: Wallflower Press.

Blackman, A. (2013). *A Virtual Love*. London: Legend Press.

Blanchot, M. (1971). *Friendship* (trans. Rottenberg, E.). California: Stanford University Press.

Blumler, J., and Gurevitch, M. (1995). *The Crisis in Public Communication*. London: Routledge.

Boellstorff, T. (2008). *Coming of Age in Second Life: An Anthropologist Explores the Virtually Human*. Princeton: Princeton University Press.

Bogost, I. (2006). *Unit Operations*. Massachusetts: MIT Press.

Bolter, J., and Grusin, R. (2000). *Remediation*. Massachusetts: MIT Press.

Boof, K. (2006). *Diary of a Lost Girl*. California: Door of Kush Multimedia.

Borges, J. (1970). *Labyrinths* (trans. Yates, D., and Irby, J.). Harmondsworth: Penguin.

Boswell, J. (1986). *The Life of Samuel Johnson*. Harmondsworth: Penguin 1986.

Bourdieu, P. (1977). *Outline of a Theory of Practice* (trans. Nice, R.). Cambridge: Cambridge University Press.

Bourdieu, P. (1986). *Distinction: A Social Critique of the Judgement of Taste* (trans. Nice, R.). London: Routledge.

Bourdieu, P. (1991). *Language and Symbolic Power* (trans. Raymond, G., and Adamson, M.). Cambridge: Polity Press.

Bourdieu, P. (2005). 'The Mystery of ministry: from particular wills to the general will' (trans. Nice, R., and Wacquant, L.) in *Pierre Bourdieu and Democratic Politics* (ed. Wacquant, L.). Cambridge: Polity Press.

Boyd, C. (2004). 'Estonia opens politics to the web' in *BBC News Interactive*, 7 May 2004.

Boyd, D. (2006). 'Friends, Friendsters, and MySpace Top 8: writing community into being on social network sites' in *First Monday* 11:12.

Boyd, D. (2008a). 'Facebook's privacy trainwreck: exposure, invasion and social convergence' in *Convergence* 14:1, 13–20.

Boyd, D. (2008b). *Taken Out of Context: American Teen Sociality in Networked Publics.* Doctoral dissertation, Berkeley, CA: University of California.

Boyd, D., and Heer, J. (2006). 'Profiles as conversation: networked identity performance on Friendster' in *Proceedings of the Hawai'i International Conference on System Sciences.*

Brady, T. (2014). 'Glasgow store to become first in Britain to replace the pound with virtual currency Bitcoin' in *Daily Mail*, 12 May 2014.

Brautigan, R. (1968). *The Pill versus the Springhill Mine Disaster.* New York: Delta Books.

Brecht, B. (1978). *Brecht on Theatre* (trans. Willett, J.). London: Methuen.

Breuer, F., and Trechsel, A. (2006). *Report for the Council of Europe: E-Voting in the 2005 local elections in Estonia.* Strasbourg: Council of Europe.

Broughton, J. (2008). *Wikipedia: The Missing Manual.* Sebastopol, CA: O'Reilly Media.

Brown, G. (2009). 'The internet is as vital as water and gas' in *The Times*, 16 June 2009.

Bruns, A. (2008). *Blogs, Wikipedia, Second Life, and Beyond: From Production to Produsage.* New York: Peter Lang.

Burke, M., Marlow, C., and Lento, T. (2010) 'Social network activity and social well-being' in *Proceedings of the 28th international conference on human factors in computing systems, Atlanta, Georgia.*

Burrell, I. (2010). 'It's not about how many pages. It's about how good they are' in *The Independent*, 20 December 2010.

Bynum, T., and Rogerson, S. (2004a). 'Ethics in the information age' in *Computer Ethics and Professional Responsibility* (ed. Bynum, T., and Rogerson, S.). Oxford: Blackwell Publishing. 1–13.

Bynum, T., and Rogerson, S. (2004b). 'Global information ethics' in *Computer Ethics and Professional Responsibility* (ed. Bynum, T., and Rogerson, S.). Oxford: Blackwell Publishing. 316–318.

Camus, A. (1975). *The Myth of Sisyphus* (trans. O'Brien, J.). Harmondsworth: Penguin.

Castronova, E. (2005). *Synthetic Worlds*. Chicago: University of Chicago Press.

Cellan-Jones, R. (2012). 'Is air leaking from the Facebook bubble?' in *BBC News Interactive*, 21 May 2012.

Chebib, N., and Sohail, R. (2011). 'The reasons social media contributed to the 2011 Egyptian revolution' in *International Journal of Business Research and Management*, 2:2, 139–162.

Chomsky, N. (1989). *Necessary Illusions*. London: Pluto Press.

Cicero, M. (2010). *Treatises on Friendship and Old Age* (trans. Shuckburgh, E.). Marston Gate: Hard Press.

Cohen, N. (2014). 'Wikipedia vs. the small screen' in *The New York Times*, 9 February 2014.

Coleman, S., and Spiller, J. (2003). 'Exploring new media effects on representative democracy' in *The Journal of Legislative Studies* 9:3, 1–16.

Coleman, S. (2005a). 'E-democracy – what's the big idea?'. Manchester: British Council Governance Team.

Coleman, S. (2005b). 'Just how risky is online voting?' in *Information Polity* 10, 95–104.

Coleman, S. (2005c). 'The lonely citizen: indirect representation in an age of networks' in *Political Communication* 22:2, 197–214.

Coleman, S. (2006). 'How the other half votes: *Big Brother* viewers and the 2005 general election' in *International Journal of Cultural Studies* 9:4, 457–479.

Coleman, S., and Blumler, J. (2009). *The Internet and Democratic Citizenship*. Cambridge: Cambridge University Press.

Condella, C. (2010). 'Why can't we be virtual friends?' in *Facebook and Philosophy* (ed. Wittkower, D.). Chicago: Carus Publishing. 111–122.

Corcoran, K. (2014). 'Twitter troll who harassed Countdown star Rachel Riley with more than 500 abusive messages given restraining order' in *Daily Mail*, 30 May 2014.

Couldry, N. (2010). 'Voice that matters' in *Media, Communication and Cultural Studies Association Conference*, London School of Economics, 6–8 January 2010.

Cummings, R. (2009). *Lazy Virtues: Teaching Writing in the Age of Wikipedia*. Nashville: Vanderbilt University Press.

Curran, J., Fenton, N., and Freedman, D. (2012). *Misunderstanding the Internet*. Abingdon: Routledge.

Curtice, J. (2009). Round table discussion in *Political Studies Association Media and Politics Group Conference*, University of Strathclyde, 5–6 November 2009.

Dalby, A. (2009). *The World and Wikipedia: How We Are Editing Reality*. Draycott: Siduri Books.

Dalton, S. (2011). 'The revolution will be streamed' in *The Times*, 25 March 2011.

Darley, A. (2000). *Visual Digital Culture*. London: Routledge.

Dawkins, R. (1989). *The Selfish Gene*. Oxford: Oxford University Press.

Day, E. (2010). 'Reality checks' in *The Observer Magazine*, 21 November 2010.

de Man, P. (1984). *The Rhetoric of Romanticism*. New York: Columbia University Press.

Dennis, J. (2014). 'All hail the keyboard radical? A new research agenda for political participation and social media' in *Political Studies Association Conference*, Manchester, 14–16 April 2014.

Derrida, J. (1981). *Dissemination* (trans. Johnson, B.). London : Athlone Press.

Derrida, J. (1987). Interview in *Criticism and Society* (ed. Salusinszky, I.). London: Methuen.

Derrida, J. (1989). *Memoires for Paul de Man* (trans. Kamuf, P.). New York: Columbia University Press.

Derrida, J. (2005). *The Politics of Friendship* (trans. Collins, G.) London: Verso.

Dickens, C. (1997). *David Copperfield*. London: Penguin.

Dixon, S. (2007). *Digital Performance: A History of New Media in Theatre, Dance Performance Art and Installation*. Cambridge, Massachusetts: MIT Press.

Downing, J. (2012). Keynote address in *Media, Communication and Cultural Studies Association Conference*, University of Bedfordshire, 11–13 January 2012.

Doyle, W., and Fraser, M. (2010). 'Facebook, surveillance and power' in *Facebook and Philosophy* (ed. Wittkower, D.). Chicago: Carus Publishing. 215–230.

Dunbar, R. (2010). 'How many friends does one person need?', paper presented to the Royal Society for the Encouragement of the Arts, Manufactures and Commerce, London, 18 February 2010.

Easton, M. (2012). 'Is our democracy moving into cyberspace?' in *BBC News Interactive*, 9 November 2012.

Egenfeldt-Nielsen, S., Smith, J. H., and Tosca, S. P. (2013). *Understanding Video Games*. New York: Routledge.

Eliot, T. S. (1925). 'The Hollow Men' in *Poems 1909–1925*. London: Faber & Faber.

Ellison, N., Steinfeld, C., and Lampe, C. (2007). 'The benefits of Facebook friends: social capital and college students' use of online social network sites' in *Journal of Mediated Communication* 12, 1143–1168.

Ernsdorff, M., and Berbec, A. (2007). 'Estonia: the short road to e-government and e-democracy', in *E-government in Europe* (ed. Nixon, P., and Koutrakou, V.). Abingdon: Routledge. 171–183.

Estonian National Electoral Committee (2007). *Parliamentary elections 2007: Statistics of e-voting*. Tallinn: Estonian National Electoral Committee.

European Commission (2005). *Online government is now a reality almost everywhere in the EU*. Brussels: European Commission, 8 March 2005.

Evans, S. (2011). 'The self and second life: a case study exploring the emergence of virtual selves' in *Reinventing Ourselves: Contemporary Concepts of Identity in Virtual Worlds* (ed. Peachey, A., and Childs, M.). London: Springer. 33–57.

Fanon, F. (1990). *The Wretched of the Earth* (trans. Farrington, C.). Harmondsworth: Penguin.

Fenton, N. (2013). 'Foreword' in *Media/Democracy: A Comparative Study* (ed. Charles, A.). Newcastle: Cambridge Scholars Publishing.

Fisk, R. (2011), 'Egypt's day of reckoning' in *The Independent*, 28 January 2011.

Fisk, R. (2014). 'Our addiction to the internet is as harmful as any drug – and what passes for comment these days is often simply foul abuse' in *The Independent*, 25 May 2014.

Fiske, J. (1987). *Television Culture*. London: Routledge.

Fletcher, M. (2011). 'Crowds salute the Facebook dreamer who led his nation' in *The Times*, 9 February 2011.

Foucault, M. (1991). *Discipline and Punish* (trans. Sheridan, A.). Harmondsworth: Penguin.

Foucault, M. (1998). *The Will to Knowledge* (trans. Hurley, R.). London: Penguin.

Fossato, F, and Lloyd, J., with Verkhovsky, A. (2008). *The Web that Failed*. Oxford: Reuters Institute for the Study of Journalism.

Frasca, G. (2007). *Play the Message: Play, Game and Videogame Rhetoric*. PhD dissertation, Copenhagen: IT University of Copenhagen.

Frau-Meigs, D. (2007). 'Convergence, internet governance and cultural diversity' in *Ambivalence towards Convergence* (ed. Storsul, T., and Stuedahl, D.). Göteborg: Nordicom. 33–53.

Freud, S. (1985). *Art and Literature* (trans. Strachey, J.). Harmondsworth: Penguin.

Fuchs, C. (2008). *Internet and Society: Social Theory in the Information Age*. New York: Routledge.

Fuchs, C. (2011). 'New media, Web 2.0 and surveillance' in *Sociology Compass* 5:2, 134–147.

Gaber, I. (2010). 'Political journalism in retreat' in *Political Studies Association Conference*, Edinburgh, 29 March – 1 April 2010.

Genvo, S. (2009). 'Understanding digital playability' in *The Video Game Theory Reader 2* (ed. Perron, B., and Wolf, M.). London: Routledge. 133–149.

Gerges, F. (2011). 'Cry freedom' in *The Mirror*, 21 February 2011.

Ghannam, J. (2011). *Social Media in the Arab World*. Washington, D. C.: Center for International Media Assistance.

Ghonim, W. (2012). *Revolution 2.0*. New York: Houghton Mifflin Harcourt Publishing.

Gibson, R., Cantijoch, M., Galandini, S. (2014). 'The effects of online civic self-help websites on civic and community engagement' in *Political Studies Association Conference*, Manchester, 14–16 April 2014.

Gibson, R., Lusoli, W., Römmele, A., and Ward, S. (2004). 'Representative democracy and the internet' in *Electronic Democracy* (ed. Gibson, R., Römmele, A., and Ward, S.). Abingdon: Routledge. 1–16.

Giles, J. (2005). 'Internet encyclopedias go head to head' in *Nature* 438, 900–901.

Glancy, R. (2014). 'Will you read this article about terms and conditions?' in *The Guardian*, 24 April 2014.

Glassman, M., Straus, J., and Shogan, C. (2010). *Social Networking and Constituent Communications*. Washington, D. C.: Congressional Research Service.

Gold, T. (2014). 'For Peaches Geldof: a gruesome grunt of synthetic grief' in *The Guardian*, 9 April 2014.

Gove, M. (2001). 'America awakes to terrorism by timetable – and the darkest national catastrophe' in *The Times*, 12 September 2001.

Greenfield, S. (2014). 'Facebook Home could change our brains' in *The Daily Telegraph*, 6 April 2013.

Greenstein, F. (1967). 'The impact of personality on politics' in *The American Political Science Review* 61:3, 629–641.

Gregg, M. (2011). *Works's Intimacy*. Cambridge: Polity Press.

Gur, O. (2010). 'Comparing social network sites and past social systems' in *Media, Communication and Cultural Studies Association Conference*, London School of Economics, 6–8 January 2010.

Habermas, J. (1989). *The Structural Transformation of the Public Sphere* (trans. Burger, T.). Cambridge: Polity Press.

Hall, J. (2001). *Online Journalism*. London: Pluto Press.

Hall, S. (1980). 'Encoding/decoding' in *Culture, Media, Language* (ed. Hall, S., Hobson, D., Lowe, A., and Willis, P.). London: Hutchinson. 128–139.

Hand, M. (2008). *Making Digital Cultures: Access, Interactivity, and Authenticity*. Aldershot: Ashgate Publishing.

Hands, J. (2011). *@ is for Activism*. London: Pluto Press.

Harding, A. (2012). 'Joseph Kony campaign under fire' in *BBC News Interactive*, 8 March 2012.

Hayles, N. (2000). 'The condition of virtuality' in *The Digital Dialectic* (ed. Lunenfeld, P.). Massachusetts: MIT Press. 68–94.

Hayward, D. (2008). 'Games and culture' in *Under the Mask: Perspectives on the Gamer*, University of Bedfordshire, 7 June 2008.

Heim, M. (1995). 'The design of virtual reality' in *Cyberspace, Cyberbodies, Cyberpunk* (ed. Featherstone, M., and Burrows, R.). London: Sage. 65–77.

Herman, E., and Chomsky, N. (2002). *Manufacturing Consent: The Political Economy of the Mass Media*. New York: Pantheon.

Hill, A. (2005). *Reality TV: Audiences and Popular Factual Television*. London: Routledge.

Hodgkinson, T. (2008). 'With friends like these ...' in *The Guardian*, 14 January 2008.

Hodson, R., and Sullivan, T. (2012). *The Social Organization of Work*. Belmont, California: Wadsworth.

Hoggard, L. (2010). 'Tough on the causes of cat crime' in *The Independent*, 26 August 2010.

Holmes, S. (2013). 'Politics is serious business: Jacques Rancière, griefing, and the re-partitioning of the (non)sensical' in *The Fibreculture Journal* 22, 152–170.

Hughes, R. (1991). *The Shock of the New*. London: BBC Books.

Hunicke, R. (2008). 'The modern age of gaming' in *Lift08*, Geneva, 6–8 February 2008.

Hyde, M. (2010). 'Cameron's plan to plunder the oeuvre of Simon Cowell' in *The Guardian*, 1 January 2010.

Jackson, D., Graham, T., and Wright, S. (2014). 'Mobilization and mundaneness: talking politics in third online spaces' in *Political Studies Association Conference*, Manchester, 14–16 April 2014.

Jackson, P. (2011). 'England riots: dangers behind false rumours' in *BBC News Interactive*, 12 August 2011.

Jameson, F. (1991). *Postmodernism, or, The Cultural Logic of Late Capitalism*. London: Verso.

Jameson, F. (2005). *Archaeologies of the Future*. London: Verso.

Jeffries, S. (2011). 'Sock puppets, twitterjacking and the art of digital fakery' in *The Guardian*, 29 September 2011.

Johnson, S. (1755). *A Dictionary of the English Language*. London: Knapton, Longman, Hitch, Hawes, Millar and Dodsley.

Johnson, S. (1951). *The Selected Letters of Samuel Johnson*. Oxford: Oxford University Press.

Johnson, S. (1971). *Rasselas, Poems, and Selected Prose*. San Francisco: Rinehart Press.

Jones, A. (2014). 'If Twitter wants to stop making losses, it needs to take a leaf out of Facebook' in *The Independent*, 8 February 2014.

Jones, J., and Salter, L. (2012). *Digital Journalism*. London: Sage.

Jones, R. (2014). 'PayPal washes its hands of bitcoin scam' in *The Guardian*, 1 March 2014.

Juul, J. (2012). *A Casual Revolution: Reinventing Video Games and Their Players*. Massachusetts: MIT Press.

Juul, J. (2013). *The Art of Failure: An Essay on the Pain of Playing Video Games*. Massachusetts: MIT Press.

Kanai, R., Bahrami, B., Roylance, R., and Rees, G. (2011). 'Online social network size is reflected in human brain structure' in *Proceedings of the Royal Society B (Biological Sciences)*, October 2011.

Kaur, N. (2007). *Big Brother: The Inside Story*. London: Virgin Books.

Keen, A. (2008). *The Cult of the Amateur*. London: Nicholas Brealey Publishing.

Kelsey, T. (2010). *Social Networking Spaces*. New York: Springer.

Khamis, S., and Vaughn, K. (2011). 'Cyberactivism in the Egyptian revolution: how civic engagement and citizen journalism tilted the balance' in *Arab Media and Society* 14.

Khoury, D. (2011) 'Social media and the revolutions: how the internet revived the Arab public sphere and digitalized activism' in *Perspectives* 2, 80–86.

Kierkegaard, S. (2006). *Fear and Trembling* (trans. Walsh, S.). Cambridge: Cambridge University Press.

Kierkegaard, S. (2008). *The Sickness unto Death* (trans. Hannay, A.). London: Penguin.

Kierkegaard, S. (2009). *Concluding Unscientific Postscript* (trans. Hannay, A.). Cambridge: Cambridge University Press.

Kilborn, R. (2003). *Staging the Real*. Manchester: Manchester University Press.

King, G. (2005). *The Spectacle of the Real*. Bristol: Intellect.

King, G., and Krzywinska, T. (2006). *Tomb Raiders and Space Invaders*. London: I. B. Tauris.

Kirkpatrick, D. (2011). *The Facebook Effect*. London: Random House.

Kitsing, M. (2013). 'The Estonian experience shows that while online voting is faster and cheaper, it hasn't increased turnout' in *Democratic Audit*, 3 October 2013.

Klimmt, C., and Hartmann, T. (2006). 'Effectance, self-efficacy and the motivation to play video games' in *Playing Video Games: Motives, Responses and Consequences* (ed. Vorderer, P., and Bryant, J.). New Jersey: Lawrence Erlbaum. 133–145.

Knowles, J., and Glennon, R. (2014). 'The virtual constituency? Twitter, local politics and public relations' in *Political Studies Association Conference*, Manchester, 14–16 April 2014.

Koc-Michalska, K., Lilleker, D. and Surowiec, P. (2013). 'The use of the web for political participation' in *Media/Democracy* (ed. Charles, A.). Newcastle: Cambridge Scholars Publishing. 81–102.

Konzack, L. (2009). 'Philosophical game design' in *The Video Game Theory Reader 2* (ed. Perron, B., and Wolf, M.). London: Routledge. 33–44.

Koskinen, I. (2007). 'The design professions in convergence' in *Ambivalence towards Convergence* (ed. Storsul, T., and Stuedahl, D.). Göteborg: Nordicom. 117–128.

Krzywinska, T. (2008). 'Reanimating H. P. Lovecraft' in *Under the Mask: Perspectives on the Gamer*, University of Bedfordshire, 7 June 2008.

Lange, D., Böhm, C., and Naumann, F. (2010). *Extracting Structured Information from Wikipedia Articles to Populate Infoboxes*. Potsdam: University of Potsdam.

LeBor, A. (2011). 'Despots beware' in *The Times*, 17 January 2011.

Li, Z. (2004). *The Potential of America's Army the Video Game as Civilian-Military Public Sphere*. MSc dissertation, Cambridge: Massachusetts Institute of Technology.

López, L. A., Brylla, C. S., and Shaw, P. (2013). 'Introduction' in *Computer Mediated Discourse across Languages* (ed. López, L. A., Brylla, C. S., and Shaw, P.). Stockholm: Stockholm University. 11–16.

Macleod, H. (2011). 'On the frontline with Syria's switched-on cyber activists' in *The Guardian*, 16 April 2011.

Madise, Ü., Vinkel, P., and Maaten, E. (2006). *Internet Voting at the Elections of Local Government Councils on October 2005*. Tallinn: Estonian National Electoral Committee.

Madise, Ü. (2007). *Internet Voting in Estonia Free and Fair Elections*. Tallinn: Estonian National Electoral Committee.

Maitles, H., and Gilchrist, I. (2005). 'We're citizens now!: the development of positive values through a democratic approach to learning' in *Journal for Critical Education Policy Studies* 3:1.

Manovich, L. (2001). *The Language of New Media*. Cambridge, MA: MIT Press.

Margaretten, M. (2014). 'The crisis in public communication and the pursuit of authenticity: an analysis of the Twitter feeds of Scottish MPs 2008–2010' in *Parliamentary Affairs* 67:2, 328–350.

Margetts, H., and John, P. (2014). 'Chaotic pluralism: collective action and social media' in *Political Studies Association Conference*, Manchester, 14–16 April 2014.

Margolis, M. (2007). 'E-government and democratic politics', in *E-government in Europe* (ed. Nixon, P., and Koutrakou, V.). Abingdon: Routledge. 1–18.

Markoff, J. (2014). 'Google's next phase in driverless cars: no brakes or steering wheel' in *The New York Times*, 27 May 2014.

Martens, T. (2007). *Internet Voting in Practice*. Tallinn: Estonian National Electoral Committee.

Marx, K. (1976). *Capital: Volume I* (trans. Fowkes, B.). Harmondsworth: Penguin.

May, P. (2010). *Virtually Dead*. Scottsdale: Poisoned Pen Press.

McGonigal, J. (2008). 'Reality is broken' in *Game Developer's Conference*, San Francisco, 22 February 2008.

McGonigal, J. (2011). *Reality is Broken: Why Games Make Us Better and How They Can Change the World*. New York: Penguin.

McLuhan, M. (2001). *Understanding Media*. London: Routledge.

McQueen, D. (1998). *Television*. London: Arnold.

Michael, B. (2013). 'Social media, identity and democracy' in *Media/Democracy* (ed. Charles, A.). Newcastle: Cambridge Scholars Publishing. 29–47.

Middleton, C. (1999). 'Ethics man' in *Business and Technology*, January 1999. 22–27.

Miller, D. (2011). *Tales from Facebook*. Cambridge: Polity Press.

Montaigne, M. de (1991). *The Complete Essays* (trans. Screech, M.). Harmondsworth: Penguin.

Morozov, E. (2011a). *The Net Delusion*. London: Allen Lane.

Morozov, E. (2011b). 'Facebook and Twitter are just places revolutionaries go' in *The Guardian*, 7 March 2011.

Moylan, T. (1986). *Demand the Impossible: Science Fiction and the Utopian Imagination*. London: Methuen.

Myers, D. (2009). 'The video game aesthetic: play as form' in *The Video Game Theory Reader 2* (ed. Perron, B., and Wolf, M.). London: Routledge. 45–63.

Narain, J. (2014). 'Violent video game is now linked to 4 teenage deaths: coroner investigates Call of Duty after suicide of boy who played it in bedroom' in *Daily Mail*, 27 May 2014.

Nature (2006). 'Britannica attacks' in *Nature* 440, 582.

Needham, C. (2004). 'The citizen as consumer: e-government in the United Kingdom and the United States' in *Electronic Democracy* (ed. Gibson, R., Römmele, A., and Ward. S.). Abingdon: Routledge. 43–69.

Neumann, P. (2004). 'Computer security and human values' in *Computer Ethics and Professional Responsibility* (ed. Bynum, T., and Rogerson, S.). Oxford: Blackwell Publishing. 208–226.

Newman, J. (2002). 'The myth of the ergodic videogame' in *Game Studies* 2:1.

Newman, J. (2013). *Videogames*. Abingdon: Routledge.

Nieborg, D. (2006). 'First person paradoxes – the logic of war in computer games' in *Game Set and Match II: On Computer Games, Advanced Geometries and Digital Technologies* (ed. Oosterhuis, K., and Feireiss, L.). Rotterdam: Episode Publishers. 107–115.

Nietzsche, F. (1969). *Thus Spoke Zarathustra* (trans. Hollingdale, R.). Harmondsworth: Penguin.

Nietzsche, F. (2008). *Human, All Too Human* (trans. Zimmern, H.). Ware: Wordsworth Editions.

Nixon, P. (2007). 'Ctrl, alt, delete: rebooting the European Union via e-government' in *E-government in Europe* (ed. Nixon, P., and Koutrakou, V.). Abingdon: Routledge. 19–32.

Nixon, P., and Koutrakou, V. (2007). 'Introduction' in *E-government in Europe* (ed. Nixon, P., and Koutrakou, V.). Abingdon: Routledge. xviii–xxviii.

Oberholzer-Gee, F., and Waldfogel, J. (2005). 'Strength in numbers: group size and political mobilisation' in *Journal of Law and Economics* 158, 73–91.

O'Loughlin, E. (2012). 'Revolution 2.0' in *The Daily Telegraph*, 13 January 2012.

Osbourne, A. (2011). 'Moscow airport bombing: why a terrorist mastermind is sending chills down spines' in *Daily Telegraph*, 29 January 2011.

OSCE (2007). *Republic of Estonia Parliamentary Elections 4 March 2007*. Warsaw: OSCE/Office for Democratic Institutions and Human Rights.

OSCE (2011). *Estonia Parliamentary Elections 6 March 2011*. Warsaw: OSCE/Office for Democratic Institutions and Human Rights.

O'Sullivan, D. (2009). *Wikipedia: A New Community of Practice?*. Farnham: Ashgate Publishing.

Orwell, G. (1951). *Animal Farm*. Harmondsworth: Penguin.

Ouellette, L. (2009). 'Take responsibility for yourself: Judge Judy and the neoliberal citizen' in *Reality TV: Remaking Television Culture* (ed. Murray, S., and Ouellette, L.). New York: New York University Press. 223–242.

Palfrey, J., and Gasser, U. (2008). *Born Digital*. New York: Basic Books.

Papacharissi, Z. (2010). *A Private Sphere: Democracy in a Digital Age*. Cambridge: Polity Press.

Parsons, R. (2014). 'Third arrest over offensive Leeds teacher murder posts' in *Yorkshire Evening Post*, 7 May 2014.

Pleace, N. (2007). 'E-government and the United Kingdom' in *E-government in Europe* (ed. Nixon, P., and Koutrakou, V.). Abingdon: Routledge. 61–74.

Pratchett, L. (2007). 'Local e-democracy in Europe: democratic x-ray as the basis for comparative analysis' in *International Conference on Direct Democracy in Latin America*, Buenos Aires, 14–15 March 2007.

Prentoulis, M. (2014). 'Responding to the Greek crisis: social media, horizontal organization and networks' in *Political Studies Association Conference*, Manchester, 14–16 April 2014.

Price, J. H. (2010). 'The new media revolution in Egypt: understanding the failures of the past and looking towards the possibilities of the future' in *Democracy & Society* 7:12. 1, 18–20.

Raab, C., and Bellamy, C. (2004). 'Electronic democracy and the mixed polity' in (eds), *Electronic Democracy* (ed. Gibson, R., Römmele, A., and Ward, S.). Abingdon: Routledge. 17–42.

Radsch, C. (2011). 'Blogosphere and social media' in *Seismic Shift: Understanding Change in the Middle East* (ed. Laipson, E.). Washington, DC: The Henry L. Stimson Center. 67–81.

Reagle, J. (2010). *Good Faith Collaboration: The Culture of Wikipedia*. Cambridge, Massachusetts: MIT Press.

Redden, J., and Witschge, T. (2010). 'A new news order?' in *New Media, Old News* (ed. Fenton, N.). London: Sage. 171–186.

Rice, J. (2009). *The Church of Facebook: How the Hyperconnected are Redefining Community*. Colorado Springs: David C. Cook.

Roberts, G. (2006). 'Orange suspends blogger over his Lefty Lexicon' in *The Independent*, 18 August 2006.

Robinson, N. (2006). 'Prescott for dummies' in *BBC News Interactive*, 5 July 2006.

Rubin, G. (2014). 'Antisocial networks: how to avoid Facebook friends and irritate people' in *The Observer*, 1 June 2014.

Sajuria, J. (2014). 'Are we bowling at all? An analysis of social capital on online networks' in *Political Studies Association Conference*, Manchester, 14–16 April 2014.

Sartre, J.-P. (1969). *Being and Nothingness* (trans. Barnes, H.). London: Methuen.

Sartre, J.-P. (2000). *Nausea* (trans. Baldick, R.). London: Penguin.

Savigny, H., and Temple, M. (2010). 'Politics marketing models: the curious incident of the dog that doesn't bark' in *Political Studies* 58:5, 1049–1064.

Schmeink, L. (2008). 'The horror of being human: dystopia, alternate realities and post-human societies in recent videogames' in *Bridges to Utopia*, University of Limerick, 3–5 July 2008.

Schmemann, S. (2001). 'Hijacked jets destroy twin towers and hit Pentagon' in *The New York Times*, 12 September 2001.

Scullion, R. (2013). 'Making it easy to resist' in *Media/Democracy* (ed. Charles, A.). Newcastle: Cambridge Scholars Publishing. 49–67.

Siibak, A. (2009). 'Constructing the self through the photo selection – visual impression management on social networking websites' in *Cyberpsychology: Journal of Psychosocial Research on Cyberspace*, 3:1.

Silverman, J. (2014). 'YouTube if you want to: camera phones, investigative journalism and social control' in *The End of Journalism Version 2.0* (ed. Charles, A.). Oxford: Peter Lang. 115–132.

Singh, A. (2011). 'Role of Twitter and Facebook in Arab Spring uprising overstated, says Hisham Matar' in *Daily Telegraph*, 11 July 2011.

Smallwood, J. (2010). 'Facebook: should parents "friend" their children?' in *BBC News Interactive*, 10 December 2010.

Soon, C., Brass, M., Heinze H.-J., and Haynes, J.-D. (2008). 'Unconscious determinants of free decisions in the human brain' in *Nature Neuroscience* 11, 543–545.

Sorice, M., and De Blasio, E. (2014). 'Radicals, rebels and maybe beyond: social movements, women's leadership and Web 2.0 in the Italian political sphere' in *Political Studies Association Conference*, Manchester, 14–16 April 2014.

Stevens, J. (2014). 'Top BBC editor brands UKIP racist and sexist on Twitter: news channel boss accused of bias hours before the election' in *Daily Mail*, 22 May 2014.

Stewart, G. (2014). 'I cant belive a war started and Wikipedia sleeps: news by online encyclopaedia' in *The End of Journalism: Version 2.0* (ed. Charles, A.). Oxford: Peter Lang. 133–151.

Storsul, T., and Stuedahl, D. (2007). 'Introduction' in *Ambivalence towards Convergence* (ed. Storsul, T., and Stuedahl, D.). Göteborg: Nordicom. 9–16.

Swanson, D., and Mancini, P. (1996). 'Patterns of modern electoral campaigning and their consequences' in *Politics, Media and Modern Democracy* (ed. Swanson, D., and Mancini, P.). Westport: Praeger. 247–276.

Talbot, M. (2007). *Media Discourse: Representation and Interaction*. Edinburgh: Edinburgh University Press.

Taylor, J. (2010). 'Kanye West follows only one – but who is Steven of Coventry?' in *The Independent*, 2 August 2010.

Tedesco, M. (2010). 'The Friendship that makes no demands' in *Facebook and Philosophy* (ed. Wittkower, D.). Chicago: Carus Publishing. 123–134.

Tincknell, E., and Raghuram, P. (2004). '*Big Brother*: reconfiguring the active audience of cultural studies?' in *Understanding Reality Television* (ed. Holmes, S., and Jermyn, D.). London: Routledge. 252–269.

Tiffin, J., and Rajasingham, L. (1995). *In Search of the Virtual Class*. Abingdon: Routledge.

Tiffin, J., and Rajasingham, L. (2003). *The Global Virtual University*. London: RoutledgeFalmer.

Travis, A., and Arthur, C. (2014). 'EU court backs right to be forgotten: Google must amend results on request' in *The Guardian*, 13 May 2014.

Trechsel, A. (2007). *Report for the Council of Europe: Internet voting in the March 2007 Parliamentary Elections in Estonia*. Strasbourg: Council of Europe.

Trost, C., and Grossmann, M. (2005). *Win the Right Way*. Berkeley: Public Policy Press.

Tumber, H., and Webster, F. (2006). *Journalists Under Fire*. London: Sage.

UN World Public Sector Report (2003). *E-Government at the Crossroads*. New York: United Nations.

Valenzuela, S., Park, N, and Kee, K. (2009). 'Is there social capital in a social network site?' in *Journal of Computer-Mediated Communication* 14, 875–901.

van Ham, P. (2001). 'The rise of the brand state: the postmodern politics of image and reputation' in *Foreign Affairs* 80:5, 2–6.

van Zoonen, L. (2004). 'Desire and resistance: *Big Brother* in the Dutch public sphere' in *Big Brother International: Formats, Critics and Publics* (ed. Mathijs, E., and Jones, J.). London: Wallflower Press. 16–24.

van Zoonen, L. (2010). 'Islam on the popular battlefield: performing politics and religion on YouTube' in *Political Studies Association Media and Politics Group Conference*, Loughborough University, 4–5 November 2010.

Varandani, S. (2014). 'Thai military says Facebook blockage was a technical problem and has been fixed' in *International Business Times*, 28 May 2014.

Vincent, J. (2014a). 'Iranian court serves "Zionist" Mark Zuckerberg with summons for breaches of privacy' in *The Independent*, 27 May 2014.

Vincent, J. (2014b). 'Facebook can work out what you're watching by listening through your smartphone' in *The Independent*, 28 May 2014.

Warf, B. (2013). *Global Geographies of the Internet*. London: Springer.

Waterfield, B., Dominiczak, P., and Squires, N. (2014). 'Merkel miffed at Barack Obama and David Cameron nuclear war game' in *The Daily Telegraph*, 25 March 2014.

Wesch, M. (2008). 'The Machine is Us/ing Us', paper presented at the US. Library of Congress, 23 June 2008.

Wilkinson, H. (2010). 'Spin is dead! Long live spin' in *Political Insight* 1:2, 45–47.

Williams, D. (2011). 'Rioters call for Gaddafi to go' in *Daily Mail*, 17 February 2011.

Williams, R. (2014). 'Rambo-style gunman has updated Facebook account while on run over triple Mountie killing' in *The Independent*, 5 June 2014.

Williams, Z. (2003). 'The final irony' in *The Guardian*, 28 June 2003.

Williamson, A. (2009). *MPs Online: Connecting with Constituents*. London: Hansard Society.

Wilson, G. (2011). 'I predict a rioter' in *The Sun*, 25 October 2011.

Withnall, A. (2014). 'UKIP councillor in homophobia and racism row over Elton John "pervert" and immigrant "scum" comments – just 48 hours after being elected' in *The Independent*, 25 May 2014.

Wright, S. (2011). 'Downing Street e-petitions: enhancing political communication and democracy?' in *Political Studies Association Conference*, London, 19–21 April 2011.

Wright, S. (2012). 'Assessing (e-)democratic innovations: democratic goods and Downing Street e-petitions' in *Journal of Information Technology & Politics* 9:4, 453–470.

Wright, T., Boria, E., and Breidenbach, P. (2002). 'Creative player actions in FPS online video games: playing Counter-Strike' in *Game Studies* 2:2.

Xenakis, A., and Macintosh, A. (2007). 'A methodology for the redesign of the electoral process to an e-electoral process' in *International Journal Electronic Governance* 1:1, 4–16.

Žižek, S. (1989). *The Sublime Object of Ideology*. London: Verso Books.

Žižek, S. (1999). 'The Cyberspace Real' in *Problemi* 37:3/4, 5–16.

Žižek, S. (2002). *Welcome to the Desert of the Real!*. London: Verso Books.

Žižek, S. (2008). *Violence*. London: Profile Books.

Žižek, S. (2013). 'Edward Snowden, Chelsea Manning and Julian Assange: our new heroes' in *The Guardian*, 3 September 2013.

Zywica, J., and Danowski, J. (2008). 'The faces of Facebookers: investigating social enhancement and social compensation hypotheses' in *Journal of Computer-Mediated Communication* 14, 1–34.

Index